FUNDAMENTALS OF GENERAL LINEAR ACOUSTICS

FUNDAMENTALS OF GENERAL LINEAR ACOUSTICS

Finn Jacobsen
Technical University of Denmark (DTU), Denmark

Peter Møller Juhl
University of Southern Denmark, Denmark

A John Wiley & Sons, Ltd., Publication

This edition first published 2013

© 2013, John Wiley & Sons Ltd

Registered office
John Wiley & Sons Ltd, The Atrium, Southern Gate, Chichester, West Sussex, PO19 8SQ, United Kingdom

For details of our global editorial offices, for customer services and for information about how to apply for permission to reuse the copyright material in this book please see our website at www.wiley.com.

Library of Congress Cataloging-in-Publication Data

Jacobsen, Finn.
  Fundamentals of general linear acoustics / Finn Jacobsen, Peter Moller Juhl.
      pages cm
  Includes bibliographical references and index.
  ISBN 978-1-118-34641-9 (hardback)
  1. Sound-waves–Transmission–Textbooks. 2. Wave-motion, Theory of–Textbooks. I. Juhl,
Peter Moller. II. Title.
  QC243.J33 2013
  620.2–dc23

                                                                    2013005223

A catalogue record for this book is available from the British Library.

ISBN: 9781118346419

Typeset in 10/12pt Times by Laserwords Private Limited, Chennai, India
Printed and bound in Malaysia by Vivar Printing Sdn Bhd

The cover picture shows the output of a circular delay-and-sum beamformer. The picture is adapted from Stewart Holmes' MSc Thesis entitled 'Spheriodal Beamforming' (Department of Electrical Engineering, Technical University of Denmark, 2012), in which beamforming with microphones in circular configurations in different baffles is analysed in detail.

Contents

About the Authors

Finn Jacobsen received an MSc in Electrical Engineering in 1974 and a PhD in Acoustics in 1981, both from the Technical University of Denmark (DTU). In 1996 he was awarded the degree of Doctor Technices by the Technical University of Denmark. In 1985 he became an Associate Professor in the Department of Acoustic Technology, DTU where he was Head of Department from 1989 to 1997. He is currently Head of Acoustic Technology, which is now a group within the Department of Electrical Engineering at DTU. His research interests include general linear acoustics, acoustic measurement techniques and signal processing, transducer technology, and statistical methods in acoustics. He has published approximately 100 papers in refereed journals and a similar amount of conference papers.

Finn Jacobsen has more than 25 years' experience with teaching acoustics at MSc level, and more than 15 years' experience with teaching fundamentals of acoustics at BSc level. He has supervised and co-supervised about 100 Masters thesis projects on acoustic topics. In the early 1990s he produced a set of lecture notes in Danish. From the end of the 1990s all lectures were given exclusively in English in the Acoustic Technology group at DTU, and Finn Jacobsen produced a completely new set of lecture notes which form the basis of this book and have frequently been updated and improved on the basis of comments from students.

Peter Møller Juhl obtained an MSc in electrical engineering from the Technical University of Denmark (DTU) in 1991 and in 1994 he received a PhD in numerical acoustics. He is currently an Associate Professor at the University of Southern Denmark, where he has had a key role in establishing the profile of acoustics in the BSc and MSc programmes in Physics and Technology. His research areas are general linear acoustics, mathematical and numerical modeling in acoustics, and source identification techniques such as beamforming and acoustic holography.

Peter Møller Juhl has 15 years of experience of teaching both basic and advanced acoustics to engineering students. Additionally he has taught physics at BSc level, and he has experience with teaching acoustics to students in the field of audiology. He has supervised approximately 50 BSc and MSc projects in acoustics. In his teaching he makes use of computer programs to visualise the theory and strengthen the understanding of the link between model, mathematical description and physical behaviour. Many of the figures in the present book have been created with these computer programs.

Preface

This book is a textbook on fundamentals of acoustic wave motion, and the topics covered by the book include duct acoustics, sound in enclosures, and sound radiation and scattering. Non-linear effects are only mentioned, and the effects of viscosity, heat conduction and mean flow are only touched upon. On the other hand, we have included classical expansions, because in our opinion there is an obvious link between technological possibilities and the *relevance* of theory. For more than ten years microphone array-based measurement techniques such as beamforming and holography, and loudspeaker array-based sound recording and reproduction techniques such as ambisonics, have made extensive use of results from classical analysis of sound fields (e.g., decompositions into spherical and cylindrical harmonics), which therefore have become more relevant than they seemed to be 30 years ago. Finally, measurements are important in acoustics, and therefore we have not only included a chapter on fundamentals of acoustic measurement techniques but also an appendix on applied signal analysis.

Acoustics is an interdisciplinary field, and throughout the world acoustic research at university level is carried out in relatively small groups, typically placed in departments focused on electrical engineering, applied physics, mechanical engineering, audiovisual engineering, or civil and environmental engineering. It has been our intention that the book should be equally accessible to readers with a background in electrical engineering, signal processing, physics and mechanical engineering.

The book is based on a number of lecture notes developed over many years and tested by numerous students. The notes have frequently been updated and improved on the basis of questions and critical comments from the students. We are grateful for the many generations of students whose comments have certainly improved the book.

We would also like to thank Jonas Brunskog for critical comments on the first draft of the book.

The book is intended to be self-contained and thus includes elementary material, but most of it is at graduate (Masters) level. It puts the emphasis on fairly detailed derivations based on the fundamental laws of physics and interpretations of the resulting formulas. In so far as possible it avoids electrical and mechanical equivalent circuits, so as to make it accessible to readers with different backgrounds. It certainly cannot replace or compete with Morse and Ingard's *Theoretical Acoustics* or Pierce's *Acoustics: An Introduction to Its Physical Principles and Applications*, but we hope that it can give the reader a good background for studying such advanced books.

Finn Jacobsen
Peter Møller Juhl

List of Symbols

a	radius of sphere [m]; acceleration [m/s^2]
A	total absorption area of room [m^2]; four-pole parameter [dimensionless]
B	bandwidth [Hz]; four-pole-parameter [kgm^{-4}s^{-1}]
c	speed of sound [m/s]
c_g	group velocity [m/s]
c_p	phase velocity [m/s]
C	four-pole parameter [kg^{-1}m^{4}s]
D	directivity function [dimensionless]; four-pole parameter [dimensionless]
E_a	sound energy [J]
$E\{\}$	expected value
f	frequency [Hz]
f_s	Schroeder frequency
f_x	probability density
F	force [N]
G	frequency domain Green's function [m^{-1} in three dimensions]
h	distance [m]
$h(t)$	impulse response
H_1	Struve function of first order
H_{xy}	frequency response
$h_m^{(1)}$	spherical Hankel function of first kind and m'th order
$h_m^{(2)}$	spherical Hankel function of second kind and m'th order
$H_m^{(1)}$	cylindrical Hankel function of first kind and m'th order
$H_m^{(2)}$	cylindrical Hankel function of second kind and m'th order
$\mathbf{I}$	sound intensity [W/m^2]
I_{inc}	incident sound intensity [W/m^2]
I_o	residual intensity [W/m^2]
I_{ref}	reference sound intensity [W/m^2]
I_x	component of sound intensity [W/m^2]
IL	insertion loss of silencer [dB]
j	imaginary unit
j_m	spherical Bessel function of m'th order
J_m	cylindrical Bessel function of m'th order

k	wavenumber [m^{-1}]
$\mathbf{k}_i$	random wavenumber vector [m^{-1}]
K	stiffness constant [N/m]
K_s	adiabatic bulk modulus [N/m^2]
l	length [m]
l_x, l_y, l_z	dimensions of rectangular room [m]
L_A	A-weighted sound pressure level [dB re p_{ref}]
L_{Aeq}	equivalent A-weighted sound pressure level [dB re p_{ref}]
L_{AE}	sound exposure level [dB re p_{ref}]
L_C	C-weighted sound pressure level [dB re p_{ref}]
L_{eq}	equivalent sound pressure level [dB re p_{ref}]
L_d	dynamic capability
L_I	sound intensity level [dB re I_{ref}]
L_p	sound pressure level [dB re p_{ref}]
L_W	sound power level [dB re P_{ref}]
L_Z	sound pressure level measured without frequency weighting [dB re p_{ref}]
M	mass [kg]; Mach number [dimensionless]; modal overlap [dimensionless]
m, n	integers [dimensionless]
$n(f)$	modal density [s]
n_m	spherical Neumann function of m'th order
$N(f)$	number of modes below f [dimensionless]
N_m	cylindrical Neumann function of m'th order
p	sound pressure [Pa]
p_+	complex amplitude of incident wave in duct
p_-	complex amplitude of reflected wave in duct
p_i	complex amplitude of incident wave in half space
p_r	complex amplitude of reflected wave in half space
$p_A(t)$	instantaneous A-weighted sound pressure [Pa]
p_{ref}	reference sound pressure [Pa]
p_{rms}	root mean square value of sound pressure [Pa]
p_0	static pressure [Pa]
P_a	sound power [W]
$P_{a,abs}$	absorbed sound power [W]
P_{ref}	reference sound power [W]
$P\{\}$	probability
q	volume velocity associated with a fictive surface [m^3/s]
Q	volume velocity of source [m^3/s]
r	radial distance in cylindrical and spherical coordinate system [m]
$\mathbf{r}$	position [m]
R	gas constant [m^{2}s^{-2}K^{-1}]; reflection factor [dimensionless]; distance [m]
R_{xy}	cross-correlation
s	standing wave ratio [dimensionless]
S	surface area [m^2]
S_{xy}	cross-spectrum
t	time [s]
T	absolute temperature [K]; averaging time [s]

T_{rev}	reverberation time [s]
TL	transmission loss of silencer [dB]
$\mathbf{u}$	particle velocity [m/s]
u_x	component of particle velocity [m/s]
U	velocity [m/s]
v	velocity [m/s]
V	volume [m^3]
w_{kin}	kinetic energy density [J/m^3]
w_{pot}	potential energy density [J/m^3]
w_{tot}	total energy density [J/m^3]
x, y, z	Cartesian coordinates [m]
Z_a	acoustic impedance [kg m^{-4}s^{-1}]
$Z_{a,r}$	acoustic radiation impedance [kg m^{-4}s^{-1}]
Z_m	mechanical impedance [kg/s]
$Z_{m,r}$	mechanical radiation impedance [kg/s]
Z_s	specific acoustic impedance [kgm^{-2}s^{-1}]
Y_s	specific acoustic admittance [m^{2}s/kg]
Y_m	mechanical admittance [s/kg]
α	absorption coefficient [dimensionless]
β	normalised admittance [dimensionless]
γ	ratio of specific heats [dimensionless]; propagation coefficient in medium [m^{-1}]
γ_z	propagation coefficient of evanescent duct mode [m^{-1}]
$\delta(\mathbf{r})$	delta function [m^{-3} in three dimensions]
δ_{mn}	Kronecker symbol ($= 0$ if $m \neq n$; $= 1$ if $m = n$)
δ_{pI}	pressure-intensity index [dB]
δ_{pIo}	residual pressure-intensity index [dB]
Δl	end correction of duct [m]
Δr	microphone separation distance [m]
Δ_{pI}	global pressure-intensity index [dB]
ΔT	sampling interval [s]
$\varepsilon\{\}$	normalised standard deviation
ε_m	Neumann symbol ($= 1$ if $m = 0$; $= 2$ if $m \geq 1$)
θ	polar angle in spherical coordinate system [radian]
λ	wavelength [m]
ξ	displacement [m]
ρ	density [kgm^{-3}]
$\sigma^2\{\}$	variance
τ	time constant [s]
φ	phase angle [radian]; azimuth angle in spherical coordinate system [radian]
ϕ	phase angle of sound pressure in a pure-tone field [radian]
ψ_m	mode shape [dimensionless]
Λ_{mn}	normalisation constant [dimensionless]

ω angular frequency [radian/s]
ω_N natural angular frequency of room [radian/s]
ω_s angular sampling frequency

$\hat{}$ indicates complex representation of a harmonic variable
$\tilde{}$ indicates an estimated quantity
$\langle \rangle_t$ indicates time averaging
$\nabla \chi$ gradient of scalar field
$\nabla \cdot \Xi$ divergence of vector field
∇^2 Laplace operator

1

Introduction

Acoustics is the science of sound, that is, wave motion in gases, liquids and solids, and the effects of such wave motion. Thus the scope of acoustics ranges from fundamental physical acoustics to, say, bioacoustics, psychoacoustics and music, and includes technical fields such as transducer technology, sound recording and reproduction, design of theatres and concert halls, and noise control. In this textbook we focus on fundamentals of wave motion in air at audible frequencies, technical fields are only touched upon, and perceptional aspects of sound are not dealt with.

Fundamentals of General Linear Acoustics, First Edition. Finn Jacobsen and Peter Møller Juhl.
© 2013 John Wiley & Sons, Ltd. Published 2013 by John Wiley & Sons, Ltd.

2

Fundamentals of Acoustic Wave Motion

2.1 Fundamental Acoustic Concepts

One of the characteristics of fluids, that is, gases and liquids, is the lack of constraints to deformation. Fluids are unable to transmit shearing forces, and therefore they react against a change of *shape* only because of inertia. On the other hand a fluid reacts against a change in its *volume* with a change of the pressure. Sound waves are compressional oscillatory disturbances that propagate in a fluid. The waves involve molecules of the fluid moving back and forth in the direction of propagation (with no net flow), accompanied by changes in the pressure, density and temperature; see Figure 2.1. The *sound pressure*, that is, the difference between the instantaneous value of the total pressure and the static pressure, is the quantity we hear. It is also much easier to measure the sound pressure than, say, the density or temperature fluctuations. Note that sound waves are *longitudinal waves* with the particles moving back and forth in the direction of propagation, unlike bending waves on a beam or waves on a stretched string, which are *transversal waves* in which the particles move back and forth in a direction perpendicular to the direction of propagation; see Figure 2.2.

In most cases of relevance for acoustics the oscillatory changes undergone by the fluid are extremely small. Under normal ambient conditions (101.3 kPa, 20°C) the density of air is $1.204 \, \text{kgm}^{-3}$. One can get an idea about the orders of magnitude of the oscillatory acoustic changes by considering the variations in air corresponding to a sound pressure level[1] of 120 dB, which is a very high sound pressure level, close to the threshold of pain. At this level the fractional pressure variations (the sound pressure relative to the static pressure) are about 2×10^{-4}, the fractional changes of the density are about 1.4×10^{-4}, the oscillatory changes of the temperature are less than 0.02°C, and the particle velocity[2] is about 50 mm/s, which at 1000 Hz corresponds to a particle displacement of less than 8 μm.

---

[1] See Chapter 4 for a definition of the sound pressure level.
[2] The concept of fluid particles refers to a macroscopic average, not to individual molecules; therefore the particle velocity can be much less than the velocity of the molecules associated with Brownian motion.

Fundamentals of General Linear Acoustics, First Edition. Finn Jacobsen and Peter Møller Juhl.
© 2013 John Wiley & Sons, Ltd. Published 2013 by John Wiley & Sons, Ltd.

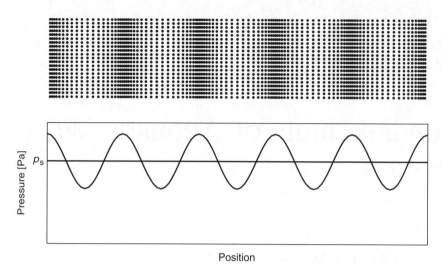

Figure 2.1 Over- and underpressure corresponding to compression and rarefaction in a sound wave

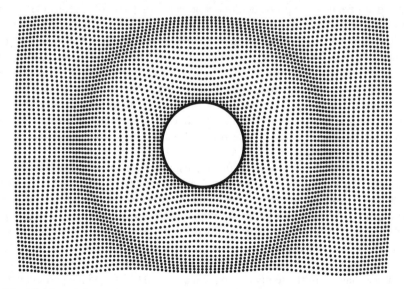

Figure 2.2 Fluid particles and compression and rarefaction in a propagating spherical sound field generated by a pulsating sphere

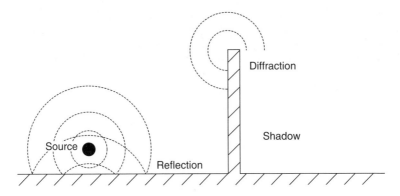

Figure 2.3   Various wave phenomena

In fact at 1000 Hz the particle displacement at the threshold of hearing is less than the diameter of a hydrogen atom![3]

Sound waves exhibit a number of phenomena that are characteristic of waves; see Figure 2.3. Waves propagating in different directions *interfere*; waves will be *reflected* by a rigid surface and more or less *absorbed* by a soft one; they will be *scattered* by small obstacles; if sources are moving a *Doppler shift* can occur; because of *diffraction* there will only partly be shadow behind a screen; and if the medium is inhomogeneous for instance because of temperature gradients the waves will be *refracted*, which means that they change direction as they propagate. The speed with which sound waves propagate in fluids is independent of the frequency, but other waves of interest in acoustics, bending waves on plates and beams, for example, are *dispersive*, which means that the speed of such waves depends on the frequency content of the waveform.

2.2   The Wave Equation

A mathematical description of the acoustic wave motion in a fluid can be obtained by combining equations that express the facts that i) mass is conserved, ii) the local longitudinal force caused by a difference in the local pressure is balanced by the inertia of the medium, and iii) sound is very nearly an adiabatic phenomenon, that is, there is no flow of heat. In what follows a homogeneous medium with no mean flow, characterised by the static pressure p_0 and the equilibrium density ρ_0 is disturbed by some mechanism.

Conservation of mass implies that a positive or negative local divergence[4] of matter must be accompanied by a corresponding change in the local density of the medium,

$$\nabla \cdot (\rho_{\mathrm{tot}} \mathbf{u}) + \frac{\partial \rho_{\mathrm{tot}}}{\partial t} = 0, \tag{2.1}$$

---

[3] At these conditions the fractional pressure variations amount to about 2.5×10^{-10}. By comparison, a change in altitude of *one metre* gives rise to a fractional change in the static pressure that is about 4 00 000 times larger, about 10^{-4}. Moreover, inside an aircraft at cruising height the static pressure is typically only 80% of the static pressure at sea level. In short, the acoustic pressure fluctuations are *extremely* small compared with commonly occurring static pressure variations.

[4] Expressed in Cartesian coordinates the divergence of the vector field Ξ is $\nabla \cdot \Xi = \frac{\partial \Xi_x}{\partial x} + \frac{\partial \Xi_y}{\partial y} + \frac{\partial \Xi_z}{\partial z}$.

where

$$\rho_{\text{tot}} = \rho_0 + \rho' \tag{2.2}$$

is the total density of the medium, which is the sum of the equilibrium value ρ_0 and the small, time-varying perturbation ρ', and $\mathbf{u}$ is the particle velocity, which is a vector.

Another fundamental equation expresses conservation of momentum. A fluid particle will move because of a gradient in the pressure,[5]

$$\nabla p_{\text{tot}} + \rho_{\text{tot}} \frac{d\mathbf{u}}{dt} = \mathbf{0}, \tag{2.3}$$

where

$$p_{\text{tot}} = p_0 + p' \tag{2.4}$$

is the total pressure, which is the sum of the static pressure p_0 and the small, time-varying *sound pressure p'*, and

$$\frac{d\mathbf{u}}{dt} = \frac{\partial\mathbf{u}}{\partial t} + \frac{\partial\mathbf{u}}{\partial x}\frac{d\xi_x}{dt} + \frac{\partial\mathbf{u}}{\partial y}\frac{d\xi_y}{dt} + \frac{\partial\mathbf{u}}{\partial z}\frac{d\xi_z}{dt} = \frac{\partial\mathbf{u}}{\partial t} + (\mathbf{u}\cdot\nabla)\mathbf{u}, \tag{2.5}$$

in which $\xi = (\xi_x, \xi_y, \xi_z)$ is the position of 'the particle', and

$$\frac{d\xi}{dt} = \frac{d(\xi_x, \xi_y, \xi_z)}{dt} = \mathbf{u}(\xi_x, \xi_y, \xi_z, t). \tag{2.6}$$

The difference between $d\mathbf{u}/dt$ and $\partial\mathbf{u}/\partial t$ is due to the fact that the particle is moving.

The third fundamental relation is due to the fact that sound in air is an adiabatic phenomenon, which means that there is no local heat exchange.[6] Under such conditions the relation between the total pressure and the total density is a power law,

$$p_{\text{tot}} = K_s \rho_{\text{tot}}^\gamma, \tag{2.7}$$

where

$$\gamma = \frac{c_p}{c_V} \tag{2.8}$$

is the ratio of specific heat at constant pressure to that at constant volume ($\simeq 1.401$ for air). Differentiating Equation (2.7) with respect to time gives

$$\left.\frac{\partial p_{\text{tot}}}{\partial t}\right|_{p_0} = \left.\frac{\partial p_{\text{tot}}}{\partial \rho_{\text{tot}}}\right|_{p_0}\left.\frac{\partial \rho_{\text{tot}}}{\partial t}\right|_{p_0} = c^2\left.\frac{\partial \rho_{\text{tot}}}{\partial t}\right|_{p_0}, \tag{2.9}$$

where we have introduced the quantity

$$c^2 = \left.\frac{\partial p_{\text{tot}}}{\partial \rho_{\text{tot}}}\right|_{p_0} = \left.K_s\gamma\rho_{\text{tot}}^{\gamma-1}\right|_{p_0} = \left.\frac{\gamma p_{\text{tot}}}{\rho_{\text{tot}}}\right|_{p_0} = \frac{\gamma p_0}{\rho_0}, \tag{2.10}$$

which as we shall see later is the square of the speed of sound ($\simeq 343\,\text{ms}^{-1}$ for air at 20°C).

---

[5] Expressed in Cartesian coordinates the gradient of the scalar field χ is $\nabla\chi = \left(\frac{\partial\chi}{\partial x}, \frac{\partial\chi}{\partial y}, \frac{\partial\chi}{\partial z}\right)$.

[6] Very near solid walls heat conduction cannot be ignored and process tends to be isothermal, which means that the temperature is constant.

Adiabatic compression

The absence of local exchange of heat implies that the entropy of the medium is constant [1, 2], and that pressure fluctuations are accompanied by density variations *and* temperature fluctuations. From Equation (2.7) it follows that

$$p_0 + p' = K_s(\rho_0 + \rho')^{\gamma},$$

which shows that

$$1 + \frac{p'}{p_0} = \left(1 + \frac{\rho'}{\rho_0}\right)^{\gamma},$$

or to the first order

$$\frac{p'}{p_0} = \frac{\gamma \rho'}{\rho_0},$$

indicating that the fractional pressure associated with adiabatic compression exceeds the fractional increase of the density (or reduction of the volume) by a factor of γ because of the increase of the temperature. The temperature fluctuations can be found in a similar manner. From the ideal gas equation,

$$p_{\text{tot}} = RT \rho_{\text{tot}},$$

where R is the gas constant ($\simeq 287 \, \text{Jkg}^{-1}\text{K}^{-1}$) and T is the ambient absolute temperature, we conclude that

$$\frac{p'}{p_0} = \frac{\rho'}{\rho_0} + \frac{T'}{T},$$

where T' is the change in temperature, or

$$T' = T\left(\frac{p'}{p_0} - \frac{\rho'}{\rho_0}\right) = T\frac{\gamma - 1}{\gamma}\frac{p'}{p_0}.$$

The observation that most acoustic phenomena involve perturbations that are several orders of magnitude smaller than the equilibrium values of the medium makes it possible to simplify the mathematical description by neglecting higher-order terms. Equation (2.1) now becomes

$$\rho_0 \nabla \cdot \mathbf{u} + \frac{\partial \rho'}{\partial t} = 0, \tag{2.11}$$

which with Equation (2.9) becomes

$$\rho_0 \nabla \cdot \mathbf{u} + \frac{1}{c^2}\frac{\partial p'}{\partial t} = 0. \tag{2.12}$$

In the same way Equation (2.3) becomes

$$\nabla p' + \rho_0 \frac{\partial \mathbf{u}}{\partial t} = \mathbf{0}. \tag{2.13}$$

Equations (2.12) and (2.13) are the *linear acoustic equations*; and Equation (2.13) is also known as Euler's equation of motion. Taking the divergence of Equation (2.13) and combining with Equation (2.12) gives

$$\nabla \cdot (\nabla p') + \rho_0 \nabla \cdot \left(\frac{\partial \mathbf{u}}{\partial t} \right) = \nabla^2 p' + \rho_0 \frac{\partial}{\partial t}(\nabla \cdot \mathbf{u}) = \nabla^2 p' - \frac{1}{c^2} \frac{\partial^2 p'}{\partial t^2} = 0, \qquad (2.14)$$

that is,

$$\nabla^2 p' = \frac{1}{c^2} \frac{\partial^2 p'}{\partial t^2}. \qquad (2.15)$$

This is the *linearised wave equation* expressed in terms of the sound pressure p'. The operator ∇^2 is the Laplacian.[7] As mentioned earlier the quantity

$$c = \sqrt{\gamma p_0 / \rho_0} = \sqrt{K_s / \rho_0} = \sqrt{\gamma R T}, \qquad (2.16)$$

where the last equation is due to the ideal gas expression, is the *speed of sound*. Note that this is proportional to the square root of the absolute temperature and independent of the static pressure. The quantity K_s is the *adiabatic bulk modulus*,

$$K_s = \rho_0 \left. \frac{\partial p_{tot}}{\partial \rho_{tot}} \right|_{p_0} = \gamma p_0 = \rho_0 c^2 \qquad (2.17)$$

($\simeq 142$ kPa for air under normal ambient conditions). It is apparent that this depends only on the static pressure, not on the temperature. Finally we can note that the equilibrium density of the medium can be written

$$\rho_0 = \frac{p_0}{RT}, \qquad (2.18)$$

which shows that it is proportional to the static pressure and inversely proportional to the absolute temperature.

Example 2.1 Adiabatic compression: Because the process of sound is adiabatic, the fractional pressure variations in a small cavity (see Figure 2.4) driven by a vibrating piston, say, a pistonphone for calibrating microphones, equal the fractional density variations multiplied by the ratio of specific heats γ. The physical explanation for the 'additional' pressure is that the pressure increase/decrease caused by the reduced/expanded volume of the cavity is accompanied by an increase/decrease of the temperature, which increases/reduces the pressure even further. The fractional variations in the density are of course identical with the fractional change of the volume (except for the sign); therefore,

$$\frac{p'}{p_0} = \gamma \frac{\rho'}{\rho} = -\gamma \frac{\Delta V}{V}.$$

---

[7] In Cartesian coordinates the Laplacian of a scalar field χ takes the form

$$\nabla^2 \chi = \left(\frac{\partial}{\partial x}, \frac{\partial}{\partial y}, \frac{\partial}{\partial z} \right) \cdot \left(\frac{\partial \chi}{\partial x}, \frac{\partial \chi}{\partial y}, \frac{\partial \chi}{\partial z} \right) = \frac{\partial^2 \chi}{\partial x^2} + \frac{\partial^2 \chi}{\partial y^2} + \frac{\partial^2 \chi}{\partial z^2}.$$

A negative value of the divergence of the pressure gradient at a certain point implies that the gradient converges towards the point, indicating a high local value. The wave equation states that this high local pressure tends to decrease at a rate that depends on the speed of sound.

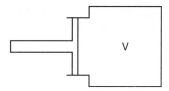

Figure 2.4 A small cavity driven by a vibrating piston

Sound in liquids

The speed of sound is much higher in liquids than in gases. For example, the speed of sound in water is about $1500\,\mathrm{ms}^{-1}$. The density of liquids is also much higher; the density of water is about $1000\,\mathrm{kgm}^{-3}$. Both the density and the speed of sound depend on the static pressure and the temperature, and there are no simple general relations corresponding to Equations (2.16) and (2.18).

The approximations that lead to the linear acoustic equations imply neglecting second- and higher-order terms in the small perturbations p' and ρ'. The resulting equations are good approximations provided that

$$p' \ll \rho_0 c^2 = \gamma p_0, \tag{2.19}$$

which, with a static pressure of the order of $100\,\mathrm{kPa}$ is seen to be the case even at a sound pressure level of $140\,\mathrm{dB}$ ($200\,\mathrm{Pa}$). The linearity of Equations (2.12), (2.13) and (2.15) is due to the absence of higher-order terms in p' and $\mathbf{u}$ in combination with the fact that $\partial/\partial x$, $\partial/\partial t$, $\partial^2/\partial x^2$ and $\partial^2/\partial t^2$ are linear operators.[8] Linearity is an extremely important property that simplifies the analysis enormously. It implies that a sinusoidal source will generate a sound field in which the pressure at all positions varies sinusoidally. It also implies linear superposition: sound waves do not interact, they simply pass through each other.[9]

The diversity of possible sound fields is of course enormous, which leads to the conclusion that we must supplement Equation (2.15) with some additional information about the sources that generate the sound field, surfaces that reflect or absorb sound, objects that scatter sound, etc. This information is known as *the boundary conditions*. The boundary conditions are often expressed in terms of the particle velocity. For example, the normal component of the particle velocity $\mathbf{u}$ is zero on a rigid surface.[10] In such cases we can

---

[8] This follows from the fact that, say, $\partial^2(p_1 + p_2)/\partial t^2 = \partial^2 p_1/\partial t^2 + \partial^2 p_2/\partial t^2$.

[9] At very high sound pressure levels, say at levels in excess of $140\,\mathrm{dB}$, the linear approximation is no longer adequate. This complicates the analysis tremendously. Nonlinear effects include waveform distortion (an overpressure travels faster than an underpressure), interaction of waves (instead of linear superposition), and formation of shock waves. It is not only the level that matters; nonlinear effects accumulate with distance. Fortunately, we can safely assume linearity under practically all circumstances encountered in daily life.

[10] Mathematicians tend to call such a boundary condition, which specifies the value of the gradient of the solution to a partial differential equation (here the sound pressure) a 'Neumann boundary condition'. The opposite extreme, a boundary condition that specifies the value of the variable of primary concern (in our case the sound pressure) directly is called a 'Dirichlet boundary condition'. A third possibility is a 'Robin boundary condition'; this condition specifies the ratio of the value of the variable of primary concern to its derivative.

make use of Euler's equation of motion, Equation (2.13), and conclude that the normal component of the pressure gradient is zero on the surface.

For later reference we shall need the Laplacian of the sound pressure expressed in the Cartesian, the cylindrical and the spherical coordinate system:[11]

$$\nabla^2 p' = \frac{\partial^2 p'}{\partial x^2} + \frac{\partial^2 p'}{\partial y^2} + \frac{\partial^2 p'}{\partial z^2}, \tag{2.20a}$$

$$\nabla^2 p' = \frac{\partial^2 p'}{\partial r^2} + \frac{1}{r}\frac{\partial p'}{\partial r} + \frac{1}{r^2}\frac{\partial^2 p'}{\partial \varphi^2} + \frac{\partial^2 p'}{\partial z^2}, \tag{2.20b}$$

$$\nabla^2 p' = \frac{\partial^2 p'}{\partial r^2} + \frac{2}{r}\frac{\partial p'}{\partial r} + \frac{1}{r^2}\frac{\partial^2 p'}{\partial \theta^2} + \frac{1}{r^2 \tan\theta}\frac{\partial p'}{\partial \theta} + \frac{1}{r^2\sin^2\theta}\frac{\partial^2 p'}{\partial \varphi^2}. \tag{2.20c}$$

Finally, we will simplify the notation from now on: instead of p' will just write p for the sound pressure, and the equilibrium density will be denoted ρ.

References

[1] P.M. Morse and K.U. Ingard: *Theoretical Acoustics*. Princeton University Press (1984). See Section 6.4.
[2] A.D. Pierce: *Acoustics. An Introduction to Its Physical Principles and Applications*. The American Institute of Physics, New York (1989). See Section 1.4.

---

[11] One example of how to derive such relationships will suffice. If the sound pressure in a spherical coordinate system depends only on r we have

$$\frac{\partial p'}{\partial x} = \frac{\partial p'}{\partial r}\frac{\partial r}{\partial x},$$

which, with

$$r = \sqrt{x^2 + y^2 + z^2},$$

becomes

$$\frac{\partial p'}{\partial x} = \frac{x}{r}\frac{\partial p'}{\partial r}.$$

Similar considerations lead to the following expression for the second-order derivative,

$$\frac{\partial^2 p'}{\partial x^2} = \frac{1}{r}\frac{\partial p'}{\partial r} + x\frac{\partial}{\partial x}\left(\frac{1}{r}\frac{\partial p'}{\partial r}\right) = \frac{1}{r}\frac{\partial p'}{\partial r} + \frac{x^2}{r}\frac{\partial}{\partial r}\left(\frac{1}{r}\frac{\partial p'}{\partial r}\right) = \frac{1}{r}\frac{\partial p'}{\partial r} + \frac{x^2}{r^2}\frac{\partial^2 p'}{\partial r^2} - \frac{x^2}{r^3}\frac{\partial p'}{\partial r}.$$

Combining Equation (2.20a) with this expression and the corresponding relations for y and z finally yields the first term of Equation (2.20c):

$$\frac{\partial^2 p'}{\partial x^2} + \frac{\partial^2 p'}{\partial y^2} + \frac{\partial^2 p'}{\partial z^2} = \frac{3}{r}\frac{\partial p'}{\partial r} + \frac{x^2 + y^2 + z^2}{r^2}\frac{\partial^2 p'}{\partial r^2} - \frac{x^2 + y^2 + z^2}{r^3}\frac{\partial p'}{\partial r} = \frac{\partial^2 p'}{\partial r^2} + \frac{2}{r}\frac{\partial p'}{\partial r}.$$

3

Simple Sound Fields

3.1 Plane Waves

The *plane wave* is a central concept in acoustics. Plane waves are waves in which any acoustic variable at a given time is a constant on any plane perpendicular to the direction of propagation. Such waves can propagate in a duct. In a limited area at a distance far from a source of sound in free space the curvature of the spherical wavefronts is negligible and the waves can be regarded as locally plane.

The plane wave is a solution to the one-dimensional wave equation,

$$\frac{\partial^2 p}{\partial x^2} = \frac{1}{c^2} \frac{\partial^2 p}{\partial t^2},$$
(3.1)

cf. Equation (2.15). It is easy to show that the expression

$$p = f_1(ct - x) + f_2(ct + x),$$
(3.2)

where f_1 and f_2 are arbitrary differentiable functions, is a solution to Equation (3.1), and it can be shown this is the general solution. Since the argument of f_1 is constant if x increases as ct it follows that the first term of this expression represents a wave that propagates undistorted and unattenuated in the positive x-direction with constant speed, c, whereas the second term represents a similar wave travelling in the opposite direction (see Figures 3.1 and 3.2).

The special case of a *harmonic* plane progressive wave is of great importance. Harmonic waves are generated by sinusoidal sources, for example a loudspeaker driven with a pure tone. A harmonic plane wave propagating in the x-direction can be written

$$p = p_1 \sin\left(\frac{\omega}{c}(ct - x) + \varphi\right) = p_1 \sin(\omega t - kx + \varphi),$$
(3.3)

where $\omega = 2\pi f$ is the angular (or radian) frequency and $k = \omega/c$ is the (angular) *wavenumber*. The quantity p_1 is known as the amplitude of the wave, and φ is a phase angle (the arbitrary value of the phase angle of the wave at the origin of the coordinate system at $t = 0$). At any position in this sound field the sound pressure varies

Fundamentals of General Linear Acoustics, First Edition. Finn Jacobsen and Peter Møller Juhl.
© 2013 John Wiley & Sons, Ltd. Published 2013 by John Wiley & Sons, Ltd.

Figure 3.1 The sound pressure in a plane wave of arbitrary waveform at two different instants of time

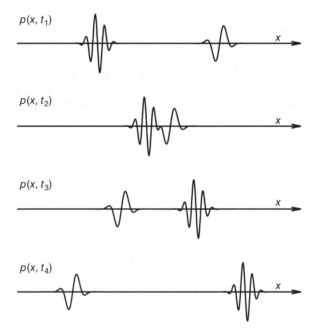

Figure 3.2 Two plane waves travelling in opposite directions are passing through each other

sinusoidally with the angular frequency ω, and at any fixed time the sound pressure varies sinusoidally with x with the spatial period

$$\lambda = \frac{c}{f} = \frac{2\pi c}{\omega} = \frac{2\pi}{k};$$
(3.4)

see Figure 3.3. The quantity λ is the *wavelength*, which is defined as the distance travelled by the wave in one cycle. Note that the wavelength is inversely proportional to the frequency. At 1000 Hz the wavelength in air is about 34 cm. In rough numbers the audible frequency range goes from 20 Hz to 20 kHz, which leads to the conclusion that acousticians are faced with wavelengths (in air) in the range from 17 m at the lowest audible frequency to 17 mm at the highest audible frequency. Since the efficiency of a radiator

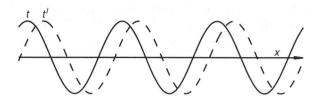

Figure 3.3   The sound pressure in a plane harmonic wave at two different instants of time

of sound or the effect of an obstacle on the sound field depends very much on its size expressed in terms of the acoustic wavelength, it can be realised that the wide frequency range is one of the challenges in acoustics. It simplifies the analysis enormously if the wavelength is very long or very short compared with typical dimensions.

Sound fields are often studied frequency by frequency. Linearity implies that a sinusoidal source with the frequency ω will generate a sound field that varies harmonically with this frequency at all positions.[1] Since the frequency is given, all that remains to be determined is the amplitude and phase at all positions. This leads to the introduction of the complex exponential representation, where the sound pressure is written as a complex function of the position multiplied with a complex exponential. The former function takes account of the amplitude and phase, and the latter describes the time dependence. Thus at any given position the sound pressure can be written as a complex function of the form[2]

$$\hat{p} = Ae^{j\omega t} = |A|e^{j\varphi}e^{j\omega t} = |A|e^{j(\omega t+\varphi)} \tag{3.5}$$

(where φ is the phase of the complex amplitude A), and the real, physical, time-varying sound pressure is the real part of the complex pressure,

$$p = \text{Re}\{\hat{p}\} = \text{Re}\{|A|e^{j(\omega t+\varphi)}\} = |A|\cos(\omega t + \varphi). \tag{3.6}$$

Since the entire sound field varies as $e^{j\omega t}$, the operator $\partial/\partial t$ can be replaced by $j\omega$ (because the derivative of $e^{j\omega t}$ with respect to time is $j\omega e^{j\omega t}$),[3] and the operator $\partial^2/\partial t^2$ can be replaced by $-\omega^2$. It follows that the one-dimensional version of Euler's equation of motion can now be written

$$j\omega\rho\hat{u}_x + \frac{\partial\hat{p}}{\partial x} = 0, \tag{3.7}$$

and the wave equation can be simplified to

$$\frac{\partial^2\hat{p}}{\partial x^2} + k^2\hat{p} = 0, \tag{3.8}$$

---

[1] If the source emitted any other signal than a sinusoidal the waveform would in the general case change with the position in the sound field, because the various frequency components would change amplitude and phase relative to each other. This explains the usefulness of harmonic analysis.

[2] Throughout this book complex variables representing harmonic signals are indicated by carets.

[3] The sign of the argument of the exponential is just a convention. The $e^{j\omega t}$ convention is common in electrical engineering, in audio engineering and in related areas of engineering acoustics. The alternative convention $e^{-j\omega t}$ is favoured by mathematicians, physicists and acousticians concerned with outdoor sound propagation. With the alternative sign convention $\partial/\partial t$ should obviously be replaced by $-j\omega$. Mathematicians and physicists also tend to prefer the symbol 'i' rather than 'j' for the imaginary unit.

which is known as the one-dimensional Helmholtz equation. See Appendix A for further details about the complex representation of harmonic signals. We note that the use of complex notation is mathematically very convenient, which will become apparent later.

Written with complex notation the equation for a plane wave that propagates in the x-direction becomes

$$\hat{p} = p_+ e^{j(\omega t - kx)}. \tag{3.9}$$

Equation (3.7) shows that the particle velocity is proportional to the gradient of the pressure. It follows that the particle velocity in the plane propagating wave given by Equation (3.9) is

$$\hat{u}_x = -\frac{1}{j\omega\rho}\frac{\partial\hat{p}}{\partial x} = \frac{k}{\omega\rho}p_+ e^{j(\omega t - kx)} = \frac{p_+}{\rho c}e^{j(\omega t - kx)} = \frac{\hat{p}}{\rho c}. \tag{3.10}$$

Thus the sound pressure and the particle velocity are in phase in a plane propagating wave, and the ratio of the sound pressure to the particle velocity is ρc, the *characteristic impedance* of the medium. As the name implies, this quantity describes an important acoustic property of the fluid, as will become apparent later. The characteristic impedance of air at 20°C and 101.3 kPa is about 413 kg·m^{-2}s^{-1}.

Example 3.1 A semi-infinite tube driven by a piston: A semi-infinite tube is driven by a piston with the vibrational velocity $U e^{j\omega t}$. Because the tube is infinite there is no reflected wave, so the sound pressure and particle velocity can be written

$$\hat{p}(x) = p_+ e^{j(\omega t - kx)}, \qquad \hat{u}_x(x) = \frac{p_+}{\rho c}e^{j(\omega t - kx)}.$$

The boundary condition at the piston implies that the particle velocity equals the vibrational velocity of the piston:

$$\hat{u}_x(0) = \frac{p_+}{\rho c}e^{j\omega t} = U e^{j\omega t}.$$

It follows that the sound pressure generated by the piston is

$$\hat{p}(x) = U\rho c\, e^{j(\omega t - kx)}.$$

The general solution to the one-dimensional Helmholtz equation is

$$\hat{p} = p_+ e^{j(\omega t - kx)} + p_- e^{j(\omega t + kx)}, \tag{3.11}$$

which can be identified as the sum of a harmonic wave that travels in the positive x-direction and a similar wave that travels in the opposite direction (cf. Equation (3.2)). The corresponding expression for the particle velocity becomes, from Equation (3.7),

$$\hat{u}_x = -\frac{1}{j\omega\rho}\frac{\partial\hat{p}}{\partial x} = \frac{k}{\omega\rho}p_+ e^{j(\omega t - kx)} - \frac{k}{\omega\rho}p_- e^{j(\omega t + kx)}$$

$$= \frac{p_+}{\rho c}e^{j(\omega t - kx)} - \frac{p_-}{\rho c}e^{j(\omega t + kx)}. \tag{3.12}$$

It can be seen that whereas $\hat{p} = \hat{u}_x \rho c$ in a plane wave that propagates in the positive x-direction, the sign is the opposite, that is, $\hat{p} = -\hat{u}_x \rho c$, in a plane wave that propagates in the negative x-direction. The reason for the change in the sign is that the particle velocity is a vector, unlike the sound pressure, so $\hat{u}_x$ is a vector component. It is also worth noting that the general relation between the sound pressure and the particle velocity in this interference field is far more complicated than in a plane propagating wave.

A plane wave that impinges on a plane rigid surface perpendicular to the direction of propagation will be reflected. This phenomenon is illustrated in Figure 3.4, which shows how an incident transient disturbance is reflected. Note that the normal component of the gradient of the pressure is identically zero on the surface for all values of t. This is a consequence of the fact that the boundary condition at the surface implies that the particle velocity must equal zero here.

However, as hinted above, it is easier to analyse the phenomenon assuming harmonic waves. In this case the sound pressure and the particle velocity are given by the general expressions (3.11) and (3.12), and our task is to determine the relation between p_+ and p_- from the boundary condition at the surface, say at $x = 0$. As mentioned, the rigid surface implies that the particle velocity must be zero here, which with Equation (3.7) leads to the conclusion that $p_- = p_+$, so the reflected wave has the same amplitude as the incident wave. Equation (3.11) now becomes

$$\hat{p} = p_+ \left(e^{j(\omega t - kx)} + e^{j(\omega t + kx)} \right) = p_+ \left(e^{-jkx} + e^{jkx} \right) e^{j\omega t} = 2p_+ \cos kx\, e^{j\omega t}, \qquad (3.13)$$

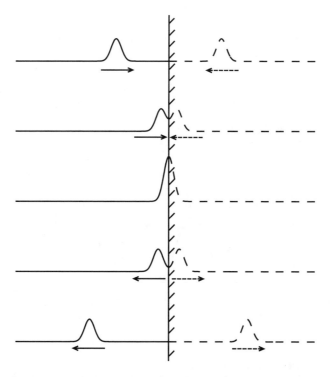

Figure 3.4 Instantaneous sound pressure in a wave that is reflected from a rigid surface at different instants of time

and Equation (3.12) becomes

$$\hat{u}_x = -j\frac{2p_+}{\rho c}\sin kx e^{j\omega t}. \tag{3.14}$$

Note that the amplitude of the sound pressure is doubled on the surface (cf. Figure 3.4). Note also the nodal[4] planes where the sound pressure is zero at $x = -\lambda/4$, $x = -3\lambda/4$, etc., and the planes where the particle velocity is zero at $x = -\lambda/2$, $x = -\lambda$, etc. The interference of the two plane waves travelling in opposite directions has produced a *standing wave pattern*, shown in Figure 3.5.

The physical explanation of the fact that the sound pressure is identically zero at a distance of a quarter of a wavelength from the reflecting plane is that the incident wave must travel a distance of half a wavelength before it returns to the same point; accordingly the incident and reflected waves are in *antiphase* (that is, 180° out of phase), and since they have the same amplitude they cancel each other. This phenomenon is called *destructive interference*. At a distance of half a wavelength from the reflecting plane the incident wave must travel one wavelength before it returns to the same point. Accordingly, the two waves are in phase and therefore the sound pressure is doubled here (*constructive interference*). The corresponding pattern for the particle velocity is different because the particle velocity is a vector.

Another interesting observation from Equations (3.13) and (3.14) is that the resulting sound pressure and particle velocity signals as functions of time at any position are 90° out of phase (since $je^{j\omega t} = e^{j(\omega t+\pi/2)}$). Otherwise expressed, if the sound pressure at a given position as a function of time is a cosine then the particle velocity at the same position is a sine. As we shall see in Chapter 6 this indicates that there is no net flow of sound energy towards the rigid surface.

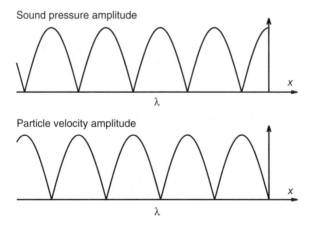

Figure 3.5 Standing wave pattern caused by reflection from a rigid surface at $x = 0$; amplitudes of the sound pressure and the particle velocity

---

[4] A *node* on, say, a vibrating string is a point that does not move, and an *antinode* is a point with maximum displacement. By analogy, points in a standing wave at which the sound pressure is identically zero are called pressure nodes. In this case the pressure nodes coincide with velocity antinodes.

> **Example 3.2 Resonances in a tube:** The standing wave phenomenon can be observed in a tube terminated by a rigid cap. When the length of the tube, l, equals an odd-numbered multiple of a quarter of a wavelength the sound pressure is zero at the input, which means that it would take very little force to drive a piston here. This is an example of an *acoustic resonance*. In this case it occurs at the frequency
>
> $$f_0 = \frac{c}{4l},$$
>
> and at odd-numbered multiples of this frequency, $3f_0$, $5f_0$, $7f_0$, etc. Note that the resonances are harmonically related. This means that if a transient mechanism excites the tube the result will be a musical sound with the fundamental frequency f_0 and overtones corresponding to odd-numbered harmonics.[5]
>
> Brass and woodwind instruments are based on standing waves in tubes. For example, closed organ pipes are tubes closed at one end and driven at the other, open end, and such pipes have only odd-numbered harmonics. See also Example 7.8.

The ratio of p_- to p_+ is the (complex) *reflection factor* R. The amplitude of this quantity describes how well the reflecting surface reflects sound. In the case of a rigid plane $R = 1$, as we have seen, which implies perfect reflection with no phase shift, but in the general case of a more or less absorbing surface R will be complex and less than unity ($|R| \leq 1$), indicating partial reflection with a phase shift at the reflection plane.

If we introduce the reflection factor in Equation (3.11) it becomes

$$\hat{p} = p_+(e^{j(\omega t - kx)} + R e^{j(\omega t + kx)}), \tag{3.15}$$

from which it can be seen that the amplitude of the sound pressure varies with position in the sound field. When the two terms in the parenthesis are in phase the sound pressure amplitude assumes its maximum value,

$$p_{max} = |p_+|(1 + |R|), \tag{3.16}$$

and when they are in antiphase the sound pressure amplitude assumes its minimum value,

$$p_{min} = |p_+|(1 - |R|). \tag{3.17}$$

The ratio of p_{max} to p_{min} is called the *standing wave ratio*,

$$s = \frac{p_{max}}{p_{min}} = \frac{1 + |R|}{1 - |R|}. \tag{3.18}$$

---

[5] A musical (or complex) tone is not a pure (sinusoidal) tone but a periodic signal (or a decaying periodic signal), usually consisting of the fundamental and a number of its harmonics, also called partials. These pure tones occur at multiples of the fundamental frequency. The n'th harmonic (or partial) is also called the $(n-1)$'th overtone, and the fundamental is the first harmonic. The relative position of a tone on a musical scale is called the *pitch* [1]. The pitch of a musical tone essentially corresponds to its fundamental frequency, which is also the distance between two adjacent harmonic components. However, pitch is a subjective phenomenon and not completely equivalent to frequency. We tend to determine the pitch on the basis of the spacing between the harmonic components, and thus we can detect the pitch of a musical tone even if the fundamental is missing.

From Equation (3.18) it follows that

$$|R| = \frac{s - 1}{s + 1},\tag{3.19}$$

which leads to the conclusion that it is possible to determine the reflection factor of a material by exposing it to normal sound incidence and measuring the standing wave ratio in the resulting interference field. See also Example 7.1.

3.2 Sound Transmission Between Fluids

When a sound wave in one fluid is incident on the boundary of another fluid, say, a sound wave in air is incident on the surface of water, it will be partly reflected and partly transmitted. For simplicity let us assume that a plane wave in fluid no 1 strikes the surface of fluid no 2 at normal incidence at $x = 0$ as shown in Figure 3.6. Anticipating a reflected wave we can write

$$\hat{p}_1(x) = p_i e^{j(\omega t - k_1 x)} + p_r e^{j(\omega t + k_1 x)} \quad \text{for } x \leq 0,\tag{3.20}$$

for fluid no 1, and

$$\hat{p}_2 = p_t e^{j(\omega t - k_2 x)} \quad \text{for } x \geq 0,\tag{3.21}$$

for fluid no 2. There are two boundary conditions at the interface: the sound pressure must be the same in fluid no 1 and in fluid no 2 (otherwise there would be a net force), and the particle velocity must be the same in fluid no 1 and in fluid no 2 (otherwise the fluids would not remain in contact). It follows that

$$p_i + p_r = p_t,\tag{3.22}$$

$$\frac{p_i - p_r}{\rho_1 c_1} = \frac{p_t}{\rho_2 c_2}.\tag{3.23}$$

Combining these equations gives

$$\frac{p_r}{p_i} = R = \frac{\rho_2 c_2 - \rho_1 c_1}{\rho_2 c_2 + \rho_1 c_1},\tag{3.24}$$

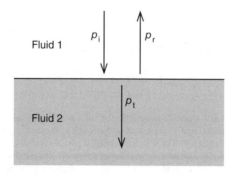

Figure 3.6 Reflection and transmission of a plane wave incident on the interface between two fluids; normal incidence

and

$$\frac{p_t}{p_i} = \frac{2\rho_2 c_2}{\rho_1 c_1 + \rho_2 c_2}, \tag{3.25}$$

which shows that the wave is almost fully reflected in phase ($R \simeq 1$) if $\rho_2 c_2 \gg \rho_1 c_1$, almost fully reflected in antiphase ($R \simeq -1$) if $\rho_2 c_2 \ll \rho_1 c_1$, and fully transmitted (not reflected at all) if $\rho_2 c_2 = \rho_1 c_1$, irrespective of the individual properties of c_1, c_2, ρ_1 and ρ_2. We can conclude that the interface to a medium with a significantly higher characteristic impedance acts as a rigid wall, whereas the interface to a medium with a significantly lower characteristic impedance acts as a very soft wall.[6] It is also interesting, and perhaps surprising, that the transmitted wave can have almost twice the amplitude of the incident wave.

Example 3.3 Water/air, normal incidence: Because of the significant difference between the characteristic impedances of air and water (the ratio is about 1: 3600) a sound wave in air that strikes a surface of water at normal incidence is almost completely reflected, and so is a sound wave that strikes the air-water interface from the water, but in the latter case the phase of the reflected wave is reversed, as shown in Figure 3.7. Compare Figures 3.4 and 3.7, and Figures 3.5 and 3.8. See also Example 6.4.

If the plane wave strikes the second medium at an oblique angle θ_1 Equation (3.20) becomes

$$\hat{p}_1 = p_i e^{j\omega t} e^{-jk_1(x\cos\theta_1 + y\sin\theta_1)} + p_r e^{j\omega t} e^{jk_1(x\cos\theta_1 - y\sin\theta_1)} \quad \text{for } x \leq 0, \tag{3.26}$$

for fluid no 1 (since it is easy to show that the reflected wave has the same angle to the normal as the incident one), whereas the transmitted wave in fluid no 2 in the general case will change direction,

$$\hat{p}_2 = p_t e^{j\omega t} e^{-jk_2(x\cos\theta_2 + y\sin\theta_2)} \quad \text{for } x \geq 0, \tag{3.27}$$

This phenomenon is called refraction; see Figure 3.9. We can now conclude that

$$(p_i + p_r)e^{-jk_1 y\sin\theta_1} = p_t e^{-jk_2 y\sin\theta_2} \tag{3.28}$$

for any value of y ('trace matching'), which leads to 'Snell's law',

$$\frac{\sin\theta_1}{\sin\theta_2} = \frac{k_2}{k_1} = \frac{c_1}{c_2}. \tag{3.29}$$

It is apparent that the transmitted wave is bent away from the normal if the speed of sound in fluid no 2 exceeds the speed of sound in the first one. It follows that when $c_2 > c_1$

---

[6] A wall with a very small impedance compared with the characteristic impedance of the given medium acts as a 'pressure-release boundary condition'.

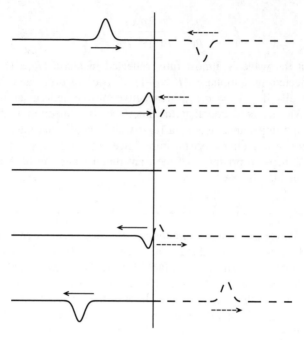

Figure 3.7 Instantaneous sound pressure in a wave that is reflected from a soft surface at different instants of time

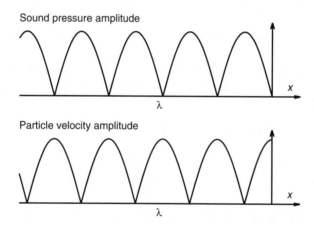

Figure 3.8 Standing wave pattern in a medium of high characteristic impedance caused by reflection from a medium of low characteristic impedance; amplitudes of the sound pressure and the particle velocity

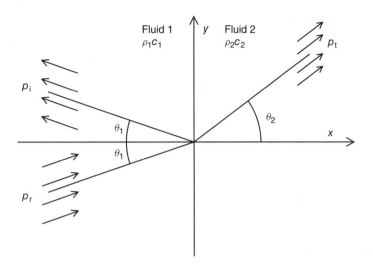

Figure 3.9 Transmission from one medium to another; oblique incidence

and θ_1 exceeds a certain 'critical angle', $\theta_c = \arcsin(c_1/c_2)$, then $\sin\theta_2 > 1$, which leads to the conclusion that

$$\cos\theta_2 = \sqrt{1 - \sin^2\theta_2} \qquad (3.30)$$

becomes purely imaginary. Thus the sound pressure in fluid no 2 can be written

$$\hat{p}_2 = p_t e^{j\omega t} e^{-\gamma x} e^{-jk_2 y \sin\theta_2} \text{ for } x \geq 0, \qquad (3.31)$$

which can be seen to be a sound field that propagates in the y-direction but decays exponentially with the distance from the interface between the two media. (Obviously we have to choose the proper sign of the square root given by Equation (3.30) – otherwise the sound pressure would grow exponentially, which is physically meaningless.) To summarise, under certain conditions there is no sound propagation away from the interface in fluid no 2 but only local wave motion in a confined region. This phenomenon is called an *evanescent wave*, and we shall meet similar phenomena both in Chapters 7 and 9. See also Example 6.5.

Example 3.4 Water/air, oblique incidence: A sound wave in air that strikes a surface of water at an angle to the normal that exceeds $\arcsin(342/1500) \simeq 0.23$ (corresponding to about 13°) is completely reflected. Conversely, sound transmitted from water to air is confined to an angle of about 13° to the normal.

From the boundary condition for the normal component of the particle velocity in the two media we can readily derive expressions for the reflected and transmitted waves relative to the incident wave,

$$\frac{p_r}{p_i} = R(\theta_1) = \frac{\rho_2 c_2 \cos\theta_1 - \rho_1 c_1 \cos\theta_2}{\rho_2 c_2 \cos\theta_1 + \rho_1 c_1 \cos\theta_2} = \frac{\dfrac{\rho_2}{\rho_1} - \sqrt{\left(\dfrac{c_1}{c_2}\right)^2 - \sin^2\theta_1}}{\dfrac{\rho_2}{\rho_1} + \sqrt{\left(\dfrac{c_1}{c_2}\right)^2 - \sin^2\theta_1}}, \tag{3.32}$$

$$\frac{p_t}{p_i} = \frac{2\rho_2 c_2 \cos\theta_1}{\rho_2 c_2 \cos\theta_1 + \rho_1 c_1 \cos\theta_2} = \frac{\dfrac{2\rho_2}{\rho_1} \cos\theta_1}{\dfrac{\rho_2}{\rho_1} \cos\theta_1 + \sqrt{\left(\dfrac{c_1}{c_2}\right)^2 - \sin^2\theta_1}}. \tag{3.33}$$

3.3 Simple Spherical Waves

The wave equation can be expressed in other coordinate systems than the Cartesian. If sound is generated by a source in an environment without reflections (which is usually referred to as a free field) it will generally be more useful to express the wave equation in a spherical coordinate system (r, θ, φ). The resulting Equation (2.20c) is obviously more complicated than Equation (2.15). However, if the source under study is spherically symmetric there can be no angular dependence, and the equation becomes quite simple,

$$\frac{\partial^2 p}{\partial r^2} + \frac{2}{r}\frac{\partial p}{\partial r} = \frac{1}{c^2}\frac{\partial^2 p}{\partial t^2}. \tag{3.34}$$

If we rewrite in the form

$$\frac{\partial^2 (rp)}{\partial r^2} = \frac{1}{c^2}\frac{\partial^2 (rp)}{\partial t^2}, \tag{3.35}$$

it becomes apparent that this equation is identical in form with the one-dimensional wave equation, Equation (3.1), although p has been replaced by rp. (It is easier to go from Equation (3.35) to Equation (3.34) than the other way.) It follows that the general solution to Equation (3.34) can be written

$$rp = f_1(ct - r) + f_2(ct + r), \tag{3.36}$$

that is,

$$p = \frac{1}{r}(f_1(ct - r) + f_2(ct + r)), \tag{3.37}$$

where f_1 and f_2 are arbitrary differentiable functions. The first term is a wave that travels outwards, away from the source (cf. the first term of Equation (3.2)). Note that the shape of the wave is preserved. However, the sound pressure is seen to decrease in inverse proportion to the distance. This is *the inverse distance law*.[7] The second term

---

[7] The inverse distance law is also known as the inverse square law because the sound intensity is inversely proportional to the square of the distance to the source. See Chapter 6.

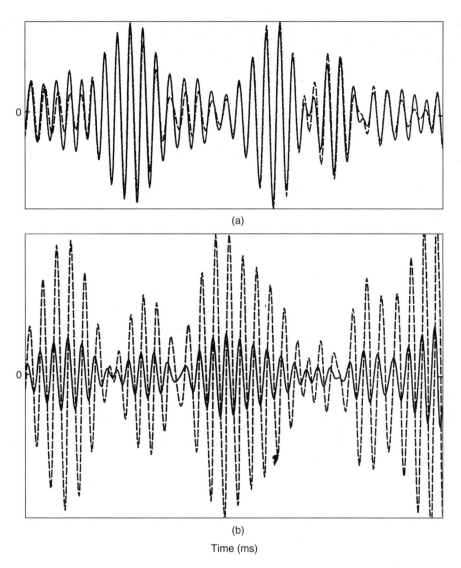

(a)

(b)

Time (ms)

Figure 3.10 (a) Measurement far from a spherical source in free space; (b) measurement close to a spherical source. Solid line, sound pressure; dashed line, particle velocity multiplied by ρc [2]

represents a converging wave, that is, a spherical wave travelling inwards. In principle such a wave could be generated by a reflecting spherical surface centred at the source, but that is a rare phenomenon indeed. Accordingly we will ignore the second term when we study sound radiation in Chapter 9. Here it suffices to mention that a diverging simple spherical sound field can be generated by a sphere that expands and contracts.

The prototype source in theoretical acoustics is a 'monopole' – an infinitely small pulsating sphere with a finite volume velocity.

A harmonic spherical wave is a solution to the Helmholtz equation

$$\frac{\partial^2 (r\hat{p})}{\partial r^2} + k^2 r \hat{p} = 0. \tag{3.38}$$

Expressed in the complex notation the diverging wave can be written

$$\hat{p} = A \frac{e^{j(\omega t - kr)}}{r}. \tag{3.39}$$

The particle velocity component in the radial direction can be calculated from Equation (3.7),

$$\hat{u}_r = -\frac{1}{j\omega\rho} \frac{\partial \hat{p}}{\partial r} = \frac{A}{\rho c} \frac{e^{j(\omega t - kr)}}{r} \left(1 + \frac{1}{jkr}\right) = \frac{\hat{p}}{\rho c} \left(1 + \frac{1}{jkr}\right). \tag{3.40}$$

Because of the spherical symmetry there are no components in the other directions. Note that far from the source[8] the sound pressure and the particle velocity are in phase and their ratio equals the characteristic impedance of the medium, just as in a plane wave. On the other hand, when $kr \ll 1$ the particle velocity is larger than $|\hat{p}|/\rho c$ and the sound pressure and the particle velocity are almost in quadrature. These are *near field* characteristics, and such a sound field is also known as a *reactive field*. See Figure 3.10.

References

[1] T.D. Rossing, F. Richard Moore and P.A. Wheeler: *The Science of Sound* (3rd edition). Addison Wesley, San Francisco, CA (2002). See Chapter 7.
[2] F. Jacobsen: A note on instantaneous and time-averaged active and reactive intensity. *Journal of Sound and Vibration* **147**, 489–496 (1991).

---

[8] In acoustics, dimensions are measured in terms of the wavelength, so that 'far from' means that $r \gg \lambda$ (or $kr \gg 1$), just as 'near' means that $r \ll \lambda$ (or $kr \ll 1$). The dimensionless quantity kr is known as the Helmholtz number.

4

Basic Acoustic Measurements

4.1 Introduction

The most important measure of sound is the root mean square sound pressure,[1] defined as

$$p_{\text{rms}} = \sqrt{\langle p^2(t)\rangle_t} = \left(\lim_{T\to\infty} \frac{1}{T} \int_0^T p^2(t)\,dt \right)^{1/2}, \tag{4.1}$$

where $\langle\ \rangle_t$ indicates averaging over time. However, as we shall see, a frequency weighting filter[2] is usually applied to the signal before the rms value is determined. Quite often such a single value does not give sufficient information about the nature of the sound, and therefore the rms sound pressure is determined in frequency bands. The resulting sound pressures are practically always compressed logarithmically and presented in decibels relative to a reference pressure.

Example 4.1 The rms-value of a sinusoidal signal: The fact that $\sin^2\omega t = \frac{1}{2}$ $(1 - \cos 2\omega t)$ and thus has a time average of $\frac{1}{2}$ leads to the conclusion that the rms value of a sinusoidal signal with the amplitude A is $A/\sqrt{2}$.

4.2 Frequency Analysis

Single frequency sound is useful for analysing acoustic phenomena, but most sounds encountered in practice have 'broadband' characteristics, which means that they cover a wide frequency range. If the sound is more or less steady, it will practically always be

---

[1] 'Root mean square' is usually abbreviated rms. The rms value is the square root of the mean square value, which is the time average of the squared signal.
[2] A filter is a device that modifies a signal by attenuating some of its frequency components.

Fundamentals of General Linear Acoustics, First Edition. Finn Jacobsen and Peter Møller Juhl.
© 2013 John Wiley & Sons, Ltd. Published 2013 by John Wiley & Sons, Ltd.

more useful to analyse it in the frequency domain than to look at the sound pressure as a function of time.[3]

Frequency (or spectral) analysis of a signal involves decomposing the signal into its spectral components. This analysis can be carried out by means of digital analysers that employ the discrete Fourier transform ('Fast Fourier Transform analysers' or 'FFT analysers'). This topic is outside the scope of this chapter; see Appendix B. Alternatively, the signal can be passed through a number of contiguous bandpass filters[4] with different centre frequencies, a 'filter bank'. The filters can have the same bandwidth or they can have constant relative bandwidth, which means that the bandwidth is a certain percentage of the centre frequency. Constant relative bandwidth corresponds to uniform resolution on a logarithmic frequency scale. Such a scale is in much better agreement with the subjective *pitch* of musical sounds than a linear scale, and therefore frequencies are often represented on a logarithmic scale in acoustics, and frequency analysis is often carried out with constant percentage filters. The most common filters in acoustics are octave filters and one-third octave filters.

An *octave*[5] is a frequency ratio of 2:1, a fundamental unit in all musical scales. Accordingly, the lower limiting frequency of an octave band is half the upper frequency limit, and the centre frequency is the geometric mean, that is,

$$f_l = f_c/2^{1/2}, \quad f_u = 2^{1/2} f_c, \quad f_c = \sqrt{f_l f_u}, \quad B = f_c(2^{1/2} - 2^{-1/2}),$$
$$(4.2a, 4.2b, 4.2c, 4.2d)$$

where f_l is the lower limiting frequency, f_c is the centre frequency, f_u is the upper limiting frequency, and B is the bandwidth. In a similar manner a one-third octave[6] is a band for which $f_u = 2^{1/3} f_l$, and

$$f_l = f_c/2^{1/6}, \quad f_u = 2^{1/6} f_c, \quad f_c = \sqrt{f_l f_u}, \quad B = f_c(2^{1/6} - 2^{-1/6}).$$
$$(4.3a, 4.3b, 4.3c, 4.3d)$$

Since $2^{10} = 1024 \simeq 10^3$ it follows that $2^{10/3} \simeq 10$ and $2^{1/3} \simeq 10^{1/10}$, that is, ten one-third octaves very nearly make a decade, and a one-third octave is almost identical with one tenth of a decade. Decades are practical, and thus in practice 'one-third octave filters' are actually one-tenth of a decade band filters. Table 4.1 gives the corresponding standardised nominal centre frequencies of octave and one-third octave filters.[7] As mentioned earlier, the human ear may respond to frequencies in the range from 20 Hz to 20 kHz, that is, a range of three decades, ten octaves or 30 one-third octaves.

---

[3] Room acoustics is an exception. Room acousticians usually prefer a description in terms of the so-called impulse response of a room.

[4] An ideal bandpass filter would allow frequency components in the passband to pass unattenuated and would completely remove frequency components outside the passband. Real filters have, of course, a certain passband ripple and a finite stopband attenuation.

[5] Musical tones an octave apart sound very similar. The diatonic scale contains seven notes per octave corresponding to the white keys on a piano keyboard; see Figure 4.1. Thus an octave spans eight notes, say, from C to C'; hence the name octave (from Latin *octo*: eight).

[6] A semitone is one twelfth of an octave on the equally tempered scale (a frequency ratio of $2^{1/12} : 1$). Since $2^{1/3} = 2^{4/12}$ it can be seen that a one-third octave is identical with four semitones or a major third (e.g., from C to E, cf. Figure 4.1). Accordingly, one-third octave band filters are called Terzfilters in German.

[7] The standardised nominal centre frequencies presented in Table 4.1 are based on the fact that the series 1.25, 1.6, 2, 2.5, 3.15, 4, 5, 6.3, 8, 10 is in reasonable agreement with $10^{n/10}$, with $n = 1, 2, \ldots 10$.

Table 4.1 Standardised nominal one-third octave and octave (**bold** characters) centre frequencies (in hertz)

20 25 **31.5** 40 50 **63** 80 100 **125** 160 200 **250** 315 400 **500** 630 800 **1000** 1250 1600 **2000** 2500 3150 **4000** 5000 6300 **8000** 10000 12500 **16000** 20000

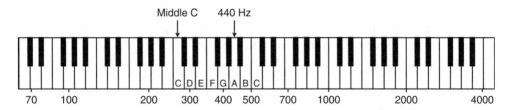

Figure 4.1 The keyboard of a small piano. The white keys from C to B correspond to the seven notes of the C major scale

An important property of the mean square value of a signal is that it can be partitioned into frequency bands. This means that if we analyse a signal in, say, one-third octave bands, the sum of the mean square values of the filtered signals equals the mean square value of the unfiltered signal. The reason is that products of different frequency components average to zero, so that all cross terms vanish; in the time domain the different filtered signals are *uncorrelated signals*. This can be illustrated by analysing a sum of two pure tones with different frequencies,

$$\left\langle \left(A \sin \omega_1 t + B \sin \omega_2 t \right)^2 \right\rangle_t = A^2 \left\langle \sin^2 \omega_1 t \right\rangle_t + B^2 \left\langle \sin^2 \omega_2 t \right\rangle_t + 2AB \left\langle \sin \omega_1 t \sin \omega_2 t \right\rangle_t$$

$$= \frac{(A^2 + B^2)}{2}. \tag{4.4}$$

Note that the mean square values of the two signals are added unless $\omega_1 = \omega_2$. The validity of this rule is not restricted to pure tones of different frequency; the mean square value of any stationary signal equals the sum of mean square values of its frequency components, which can be determined with a parallel bank of contiguous filters. Thus

$$p_{\text{rms}}^2 = \sum_i p_{\text{rms},i}^2, \tag{4.5}$$

where $p_{\text{rms},i}$ is the rms value of the output of the i'th filter. Equation (4.5) is known as Parseval's formula.

Random noise

Many generators of sound produce *noise* rather than pure tones. Whereas pure tones and other periodic signals are deterministic, *noise* is a stochastic or random phenomenon. *Stationary* noise is a stochastic signal with statistical properties that do not change with time.

White noise is stationary noise with a flat power spectral density, that is, constant mean square value per hertz. The term white noise is an analogy to white light. When white noise is passed through a bandpass filter, the mean square of the output signal is proportional to the bandwidth of the filter. It follows that when white noise is analysed with constant percentage filters, the mean square of the output is proportional to the centre frequency of the filter. For example, if white noise is analysed with a bank of octave band filters, the mean square values of the output signals of two adjacent filters differ by a factor of two.

Pink noise is stationary noise with constant mean square value in bands with constant relative width, e.g., octave bands. Thus compared with white noise low frequencies are emphasised; hence the name pink noise, which is an analogy to an optical phenomenon. It follows that the mean square value of a given pink noise signal in octave bands is three times larger than the mean square value of the noise in one-third octave bands.

See Appendix B for more about random noise.

Example 4.2 Detection of a pure tone in noise: The fact that noise, unlike periodic signals, has a finite power spectral density (mean square value *per hertz*) implies that one can detect a pure tone in noise irrespective of the signal-to-noise ratio by analysing with sufficiently fine spectral resolution: As the bandwidth is reduced, less and less noise passes through the filter, and the tone will emerge. Compared with filter bank analysers FFT analysers have the advantage that the spectral resolution can be varied over a wide range (see Appendix B); therefore FFT analysers are particular suitable for detecting tones in noise.

When several independent sources of noise are present at the same time the mean square sound pressures generated by the individual sources are additive. This is due to the fact that independent sources generate uncorrelated signals, that is, signals whose instantaneous product average to zero; therefore the cross terms vanish:

$$\left\langle \left(p_1(t) + p_2(t) \right)^2 \right\rangle_t = \left\langle p_1^2(t) \right\rangle_t + \left\langle p_2^2(t) \right\rangle_t + 2\left\langle p_1(t)\, p_2(t) \right\rangle_t = \left\langle p_1^2(t) \right\rangle_t + \left\langle p_2^2(t) \right\rangle_t. \quad (4.6)$$

It follows that

$$p_{\mathrm{rms,tot}}^2 = \sum_i p_{\mathrm{rms},i}^2. \quad (4.7)$$

Note the similarity between Equations (4.5) and (4.7). It is of enormous practical importance that the mean square values of uncorrelated signals are additive, because signals generated by different mechanisms are invariably uncorrelated. Almost all signals that occur in real life are mutually uncorrelated.

Example 4.3 Stationary independent signals: Equation (4.7) leads to the conclusion that the mean square sound pressure generated by a crowd of noisy people in a room is proportional to the number of people (provided that each person makes a given amount of noise, independently of the others). Thus the rms value of the sound pressure in the room is proportional to the *square root* of the number of people.

Example 4.4 Compensation for background noise: Consider the case where the rms sound pressure generated by a source of noise is to be measured in the presence of background noise that cannot be turned off. It follows from Equation (4.7) that it is possible to correct the measurement for the influence of the stationary background noise, at least in principle; one simply subtracts the mean square value of the background noise from the total mean square pressure. For this to work in practice the background noise must not be too strong, though, and it is absolutely necessary that it is completely stationary.

4.3 Levels and Decibels

The human auditory system can cope with sound pressure variations over a range of more than a million times. Because of this wide range, the sound pressure and other acoustic quantities are usually measured on a logarithmic scale. An additional reason is that the subjective impression of how loud noise sounds correlates much better with a logarithmic measure of the sound pressure than with the sound pressure itself. The unit is the *decibel*,[8] abbreviated dB, which is a relative measure, requiring a reference quantity. The results are called *levels*. The sound pressure level (sometimes abbreviated as SPL) is defined as

$$L_p = 10 \log_{10} \frac{p_{\text{rms}}^2}{p_{\text{ref}}^2} = 20 \log_{10} \frac{p_{\text{rms}}}{p_{\text{ref}}}, \qquad (4.8)$$

where p_{ref} is the reference sound pressure, and $\log_{10}$ is the base 10 logarithm, henceforth written log. The reference sound pressure is $20\,\mu\text{Pa}$ for sound in air, corresponding roughly to the lowest audible sound at $1\,\text{kHz}$.[9] Some typical sound pressure levels are given in Figure 4.2.

---

[8] As the name implies, the decibel is one tenth of a bel. However, the bel is rarely used today. The use of decibels rather than bels is probably due to the fact that most sound pressure levels encountered in practice take values between 0 and 120 when measured in decibels, as can be seen in Figure 4.2. Another reason might be that to be audible, the change of the level of a given (broadband) sound must be of the order of one decibel.

[9] For sound in other fluids than atmospheric air (water, for example) the reference sound pressure is $1\,\mu\text{Pa}$. To avoid possible confusion it may be advisable to state the reference sound pressure explicitly, e.g., 'the sound pressure level is $77\,\text{dB}$ re $20\,\mu\text{Pa}$.'

Figure 4.2 Typical sound pressure levels. (Copyright © Brüel & Kjær.)

The fact that the mean square sound pressures of independent sources are additive (cf. Equation (4.7)) leads to the conclusion that the levels of such sources are combined as follows:

$$L_{p,\text{tot}} = 10 \log \left(\sum_i 10^{0.1 L_{p,i}} \right). \tag{4.9}$$

Another consequence of Equation (4.7) is that one can correct a measurement of the sound pressure level generated by a source for the influence of steady background noise as follows:

$$L_{p,\text{source}} = 10 \log \left(10^{0.1 L_{p,\text{tot}}} - 10^{0.1 L_{p,\text{background}}} \right). \tag{4.10}$$

This corresponds to subtracting the mean square sound pressure of the background noise from the total mean square sound pressure as described in Example 4.4. However, since all measurements are subject to random errors, the result of the correction will be reliable only if the background level is at least, say, 3 dB below the total sound pressure level. If the background noise is more than 10 dB below the total level the correction is less than 0.5 dB.

Example 4.5 The inverse distance law expressed in decibels: Expressed in terms of sound pressure levels the inverse distance law states that the level decreases by 6 dB when the distance to the source is doubled.

Example 4.6 The combined sound pressure level of two independent sources: When each of two independent sources in the absence of the other generates a sound pressure level of 70 dB at a certain point, the resulting sound pressure level is 73 dB (**not** 140 dB!), because $10 \log 2 \simeq 3$. If one source creates a sound pressure level of 65 dB and the other a sound pressure level of 59 dB, the total level is $10 \log(10^{6.5} + 10^{5.9}) \simeq 66$ dB.

Example 4.7 The combined sound pressure level of two coherent pure tone sources: When two sinusoidal sources emit pure tones of the same frequency they create an interference field, and depending on the phase difference the total sound pressure amplitude at a given position will assume a value between the sum of the two amplitudes and the difference:

$$||A| - |B|| \leq |Ae^{j\omega t} + Be^{j\omega t}| = |A + B| = ||A|e^{j\varphi_A} + |B|e^{j\varphi_B}| \leq |A| + |B|.$$

For example, if two pure tone sources of the same frequency each generates a sound pressure level of 70 dB in the absence of the other source then the total sound pressure level can be anywhere between 76 dB (constructive interference) and $-\infty$ dB (destructive interference). Note that Equations (4.7) and (4.9) do **not** apply in this case because the signals are not uncorrelated. See also Figure A1 in Appendix A.

Other first-order acoustic quantities, for example the particle velocity, are also often measured on a logarithmic scale. The reference velocity is $1\,\mathrm{nm/s} = 10^{-9}\,\mathrm{m/s}$.[10] This reference is also used in measurements of the vibratory velocities of vibrating structures.

The acoustic second-order quantities sound intensity and sound power, defined in Chapter 6, are also measured on a logarithmic scale. The sound intensity level is

$$L_I = 10\log\frac{|I|}{I_{\text{ref}}}, \tag{4.11}$$

where I is the intensity and $I_{\text{ref}} = 1\,\mathrm{pWm^{-2}} = 10^{-12}\,\mathrm{Wm^{-2}}$,[11] and the sound power level is

$$L_W = 10\log\frac{P_a}{P_{\text{ref}}}, \tag{4.12}$$

where P_a is the sound power and $P_{\text{ref}} = 1\,\mathrm{pW}$. Note that levels of linear quantities (sound pressure, particle velocity, particle displacement) are defined as 20 times the logarithm of the ratio of the rms value to a reference value, whereas levels of second-order (quadratic) quantities are defined as *ten* times the logarithm, in agreement with the fact that if the linear quantities are doubled then quantities of second order are quadrupled.

> **Example 4.8 White noise in one-third octave bands expressed in decibels:** It follows from the constant spectral density of white noise that when such a signal is analysed in one-third octave bands, the level increases 1 dB from one band to the next $(10\log(2^{1/3}) \simeq 10\log(10^{1/10}) = 1\,\mathrm{dB})$.

4.4 Noise Measurement Techniques and Instrumentation

A sound level meter is an instrument designed to measure sound pressure levels. Today such instruments can be anything from simple devices to advanced digital analysers. Figure 4.3 shows a block diagram of a simple sound level meter. The microphone converts the sound pressure to an electrical signal, which is amplified and passes through various filters. After this the signal is squared and averaged with a detector, and the result is finally converted to decibels and shown on a display. In the following a very brief description of such an instrument will be given; see, e.g., References [1, 2] for further details.

The most commonly used microphones for this purpose are condenser microphones, which are more stable and accurate than other types of microphone. The diaphragm of

---

[10] The prefix n (for 'nano') represents a factor of 10^{-9}.
[11] The prefix p (for 'pico') represents a factor of 10^{-12}.

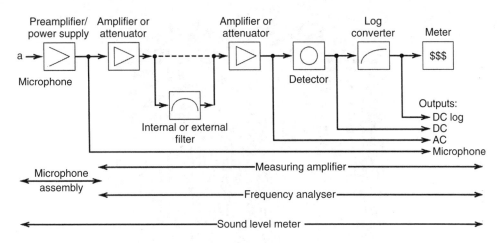

Figure 4.3   A sound level meter [1] (copyright © John Wiley & Sons Ltd)

a condenser microphone is a very thin, highly tensioned foil. Inside the housing of the microphone cartridge is the other part of the capacitor, the back plate, placed very close to the diaphragm (see Figure 4.4). The capacitor is electrically charged, either by an external voltage on the back plate or (in the case of prepolarised *electret* microphones) by properties of the diaphragm or the back plate. When the diaphragm moves in response to the sound pressure, the capacitance changes, and this produces an electrical voltage proportional to the instantaneous sound pressure.

The microphone should be as small as possible so as not to disturb the sound field. However, this is in conflict with the requirement of a high sensitivity and a low inherent noise level, and, as a compromise, typical measurement microphones are '$^1/_2$-inch' microphones with a diameter of about 13 mm. At low frequencies, say below 1 kHz, such a microphone is much smaller than the wavelength and does not disturb the sound field appreciably. In this frequency range the microphone is *omnidirectional* as of course it should be since the sound pressure is a scalar and has no direction. However, from a few kilohertz and upwards the size of the microphone is no longer negligible compared with the wavelength, and therefore it is no longer omnidirectional, which means that its response varies with the nature of the sound field; see Figure 4.5.

One can design condenser microphones to have a flat response in as wide a frequency range as possible under specified sound field conditions. For example, 'free-field' microphones are designed to have a flat response for axial incidence (see Figure 4.6), and such microphones should therefore be pointed towards the source. 'Random-incidence' microphones are designed for measurements in a diffuse sound field where sound is arriving from all directions, and 'pressure field' microphones are intended for measurements in small cavities.

The sensitivity of the human auditory system varies significantly with the frequency in a way that changes with the level. In particular the human ear is, at low levels, much less sensitive to low frequencies than to medium frequencies. This is the background for the standardised frequency weighting filters shown in Figure 4.7. The original intention was to simulate a human ear at various levels, but it has long ago been realised that the human

Figure 4.4 A condenser microphone [4] (copyright © Brüel & Kjær)

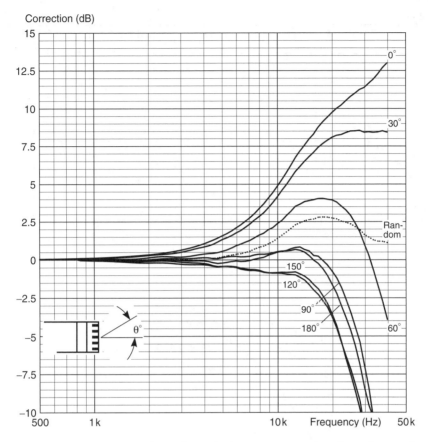

Figure 4.5 The 'free-field correction' of a typical measurement microphone for sound coming from various directions. The free-field correction is the fractional increase of the sound pressure (usually expressed in dB) caused by the presence of the microphone in the sound field [4] (copyright © Brüel & Kjær)

auditory system is far more complicated than implied by such simple weighting curves, and B- and D-weighting filters are little used today. On the other hand the A-weighted sound pressure level is the most widely used single-value measure of sound, because the A-weighted sound pressure level correlates in general much better with the subjective effect of noise than measurements of the sound pressure level with a flat frequency response. C-weighting, which is essentially flat in the audible frequency range, is sometimes used in combination with A-weighting, because a large difference between the A-weighted level and the C-weighted level is a clear indication of a prominent content of low frequency noise. The results of measurements of the A- and C-weighted sound pressure level are denoted L_A and L_C respectively, and the unit is dB.[12] If no weighting filter is applied (Z-weighting), the level is sometimes denoted L_Z.

---

[12] In practice the unit is often written dB (A) and dB (C), respectively.

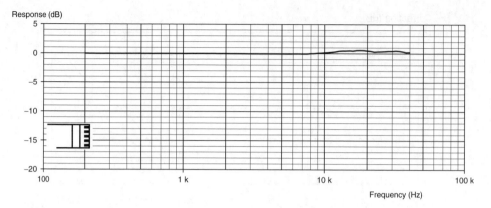

Figure 4.6   Free-field response of a microphone of the 'free-field' type at axial incidence [4] (copyright © Brüel & Kjær)

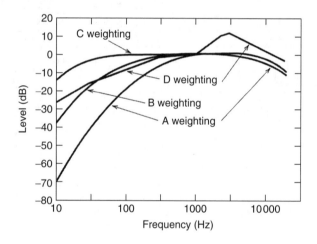

Figure 4.7   Standardised frequency weighting curves [2] (copyright © John Wiley & Sons Ltd)

In the measurement instrument the frequency weighting filter is followed by a squaring device, a lowpass filter that smooths out the instantaneous fluctuations, and a logarithmic converter. The lowpass filter corresponds to applying a time weighting function. The most common time weighting in sound level meters is exponential, which implies that the squared signal is smoothed with a decaying exponential so that recent data are given more weight than older data:

$$L_p(t) = 10 \log \left(\left(\frac{1}{\tau} \int_{-\infty}^{t} p^2(u) \, e^{-(t-u)/\tau} \, du \right) \Big/ p_{\text{ref}}^2 \right). \qquad (4.13)$$

Two values of the time constant τ are standardised: S (for 'slow') corresponds to a time constant of 1 s, and F (for 'fast') is exponential averaging with a time constant of 125 ms.

The alternative to exponential averaging is linear (or integrating) averaging, in which the mean square sound pressure is weighted uniformly during the integration. The equivalent sound pressure level is defined as

$$L_{eq} = 10 \log \left(\left(\frac{1}{t_2 - t_1} \int_{t_1}^{t_2} p^2(t) \, dt \right) \Big/ p_{ref}^2 \right). \qquad (4.14)$$

Measurements of random noise with a finite integration time are subject to random errors that depend on the bandwidth of the signal and on the integration time. It can be shown that the variance of the measurement result is inversely proportional to the product of the bandwidth and the integration time.[13]

As can be seen by comparing with Equations (4.1) and (4.14), the equivalent sound pressure level is just the sound pressure level corresponding to the rms sound pressure determined with a specified integration period. The A-weighted equivalent sound pressure level L_{Aeq} is the level corresponding to a similar time integral of the A-weighted instantaneous sound pressure. Sometimes the quantity is written $L_{Aeq,T}$ where T is the integration time.

Whereas exponential averaging corresponds to a running average and thus gives a (smoothed) measure of the sound at any instant of time, the equivalent sound pressure level (with or without A-weighting) can be used for characterising the total effect of fluctuating noise, for example noise from road traffic. Typical values of T are 30 s for measurement of noise from technical installations, 8 h for noise in a working environment and 24 h for traffic noise.

Sometimes it is useful to analyse noise signals in one-third octave bands, cf. Section 4.2. From Equation (4.5) it can be seen that the total sound pressure level can be calculated from the levels in the individual one-third octave bands, L_i, as follows:

$$L_Z = 10 \log \left(\sum_i 10^{0.1 L_i} \right). \qquad (4.15)$$

In a similar manner one can calculate the A-weighted sound pressure level from the one-third octave band values and the attenuation data given in Table 4.2,

$$L_A = 10 \log \left(\sum_i 10^{0.1(L_i + K_i)} \right), \qquad (4.16)$$

where K_i is the relative response of the A-weighting filter (in dB) in the i'th band.

Example 4.9 A-weighted level from one-third octave values: A source gives rise to the following one-third octave band values of the sound pressure level at a

---

[13] In the literature reference is sometimes made to the equivalent integration time of exponential detectors. This is two times the time constant (e.g., 250 ms for 'F'), because a measurement of random noise with an exponential detector with a time constant of τ has the same statistical uncertainty as a measurement with linear averaging over a period of 2τ [3].

certain point, 52 dB, 68 dB, 76 dB, 71 dB, 54 dB in the bands with centre frequencies 315 Hz, 400 Hz, 500 Hz, 630 Hz and 800 Hz, respectively, and less than 50 dB in all the other bands. It follows that

$$L_Z \simeq 10 \log \left(10^{5.2} + 10^{6.8} + 10^{7.6} + 10^{7.1} + 10^{5.4}\right) \simeq 77.7 \text{ dB},$$

and

$$L_A \simeq 10 \log \left(10^{(5.2-0.66)} + 10^{(6.8-0.48)} + 10^{(7.6-0.32)} + 10^{(7.1-0.19)} + 10^{(5.4-0.08)}\right)$$

$$\simeq 74.7 \text{ dB}.$$

Noise that changes its level in a regular manner is called *intermittent noise*. Such noise could for example be generated by machinery that operates in cycles. If the noise occurs at several steady levels, the equivalent sound pressure level can be calculated from the formula

$$L_{\text{eq},T} = 10 \log \left(\sum_i \frac{t_i}{T} 10^{0.1 L_i}\right). \tag{4.17}$$

This corresponds to adding the mean square values with a weighting that reflects the relative duration of each level.

Example 4.10 A-weighted level of intermittent noise: The A-weighted sound pressure level at a given position in an industrial hall changes periodically between 84 dB in intervals of 15 minutes, 95 dB in intervals of 5 minutes and 71 dB in intervals of 20 minutes. From Equation (4.17) it follows that the equivalent sound pressure level over a working day is

$$L_{\text{Aeq}} = 10 \log \left(\frac{15}{40} 10^{8.4} + \frac{5}{40} 10^{9.5} + \frac{20}{40} 10^{7.1}\right) \simeq 87.0 \text{ dB}.$$

Most sound level meters have also a peak detector for determining the highest absolute value of the instantaneous sound pressure (without filters and without time weighting), p_{peak}. The *peak level* is calculated from this value and Equation (4.8) in the usual manner, that is,

$$L_{\text{peak}} = 20 \log \frac{p_{\text{peak}}}{p_{\text{ref}}}. \tag{4.18}$$

Table 4.2 The response of standard A- and C-weighting filters in one-third octave bands

Centre frequency (Hz)	A-weighting (dB)	C-weighting (dB)
8	−77.8	−20.0
10	−70.4	−14.3
12.5	−63.4	−11.2
16	−56.7	−8.5
20	−50.5	−6.2
25	−44.7	−4.4
31.5	−39.4	−3.0
40	−34.6	−2.0
50	−30.2	−1.3
63	−26.2	−0.8
80	−22.5	−0.5
100	−19.1	−0.3
125	−16.1	−0.2
160	−13.4	−0.1
200	−10.9	0.0
250	−8.6	0.0
315	−6.6	0.0
400	−4.8	0.0
500	−3.2	0.0
630	−1.9	0.0
800	−0.8	0.0
1000	0.0	0.0
1250	0.6	0.0
1600	1.0	−0.1
2000	1.2	−0.2
2500	1.3	−0.3
3150	1.2	−0.5
4000	1.0	−0.8
5000	0.5	−1.3
6300	−0.1	−2.0
8000	−1.1	−3.0
10000	−2.5	−4.4
12500	−4.3	−6.2
16000	−6.6	−8.5
20000	−9.3	−11.2

Example 4.11 Crest factor of a sinusoidal signal: The *crest factor* of a signal is the ratio of its peak value to the rms value (sometimes expressed in dB). From Example 4.1 it follows that the crest factor of a pure tone signal is $\sqrt{2}$ or 3 dB.

The sound exposure level (sometimes abbreviated SEL) is closely related to L_{Aeq}, but instead of dividing the time integral of the squared A-weighted instantaneous sound pressure by the actual integration time one divides by $t_0 = 1$ s. Thus the sound exposure level is a measure of the total energy[14] of the noise, normalised to 1 s:

$$L_{\text{AE}} = 10 \log \left(\left(\frac{1}{t_0} \int_{t_1}^{t_2} p_{\text{A}}^2 (t) \, dt \right) / p_{\text{ref}}^2 \right) \tag{4.19}$$

This quantity is used for measuring the total energy of a 'noise event' (say, a hammer blow or the take off of an aircraft), independently of its duration. Evidently the measurement interval should encompass the entire event.

Example 4.12 Equivalent level of a noise event: It is clear from Equations (4.14) and (4.19) that $L_{\text{Aeq},T}$ of a noise event of finite duration decreases with the logarithm of T if the T exceeds its duration:

$$L_{\text{Aeq},T} = 10 \log \left(\left(\frac{1}{T} \int_{-\infty}^{\infty} p_{\text{A}}^2 (t) \, dt \right) / p_{\text{ref}}^2 \right) = L_{\text{AE}} - 10 \log \frac{T}{t_0}.$$

Example 4.13 Sound exposure level of a series of noise events: If n identical noise events each with a sound exposure level of L_{AE} occur within a period of T (e.g., one working day) then the A-weighted equivalent sound pressure level is

$$L_{\text{Aeq},T} = L_{\text{AE}} + 10 \log n - 10 \log \frac{T}{t_0},$$

because the integrals of the squared signals are additive; cf. Equation (4.7).[15]

References

[1] P.V. Brüel, J. Pope and H.K. Zaveri: Introduction to acoustical measurement and instrumentation. Chapter 154 in *Encyclopedia of Acoustics*, ed. M.J. Crocker. John Wiley & Sons, New York (1997).
[2] R.W. Krug: Sound level meters. Chapter 155 in *Encyclopedia of Acoustics*, ed. M.J. Crocker, John Wiley & Sons, New York (1997).
[3] J. Pope: Analyzers. Chapter 107 in *Handbook of Acoustics*, ed. M.J. Crocker. John Wiley & Sons, New York (1998).
[4] Anon.: *Microphone Handbook*. Brüel & Kjær, Nærum (1996).

---

[14] In signal analysis it is customary to use the term 'energy' in the sense of the integral of the square of a signal, without regard to its units. This should not be confused with the potential energy density of the sound field introduced in Chapter 6.

[15] Strictly speaking this requires that the instantaneous product of the 'event' and any of its time shifted versions time average to zero. In practice this will always be the case.

5

The Concept of Impedance

By definition an *impedance* is the ratio of the complex amplitudes of two signals representing cause and effect, for example the ratio of an AC voltage across a part of an electric circuit to the corresponding current, the ratio of a mechanical force to the resulting vibrational velocity, or the ratio of the sound pressure to the particle velocity. The term has been coined from the verb 'impede' (obstruct, hinder), indicating that it is a measure of the opposition to the flow of current, and so on. The reciprocal of the impedance is the *admittance*, coined from the verb 'admit' and indicating lack of such opposition. Note that these concepts require complex representation of harmonic signals; it makes no sense to divide, say, the instantaneous sound pressure with the instantaneous particle velocity. There is no simple way of describing properties corresponding to a complex value of the impedance without the use of complex notation.

5.1 Mechanical Impedance

The mechanical impedance is perhaps simpler to understand than the other impedance concepts we encounter in acoustics, since it is intuitively clear that it takes a certain vibratory force to generate mechanical vibrations. The mechanical impedance of a structure at a given point is the ratio of the complex amplitude of a harmonic point force acting on the structure, $\hat{F}$, to the complex amplitude of the resulting vibratory velocity at the same point, $\hat{v}$,

$$Z_m = \frac{\hat{F}}{\hat{v}}. \tag{5.1}$$

The unit is kg/s. The mechanical admittance is the reciprocal of the mechanical impedance,

$$Y_m = \frac{\hat{v}}{\hat{F}}. \tag{5.2}$$

This quantity is also known as the mobility. The unit is s/kg.

Fundamentals of General Linear Acoustics, First Edition. Finn Jacobsen and Peter Møller Juhl.
© 2013 John Wiley & Sons, Ltd. Published 2013 by John Wiley & Sons, Ltd.

Example 5.1 Mechanical impedance of a mass: It takes a force of $F = a \cdot M$ to set a mass M into the acceleration a (Newton's second law of motion), and this is also the case if the force is harmonic; therefore the mechanical impedance of the mass is

$$Z_{\mathrm{m}} = \frac{\hat{F}}{\hat{v}} = \frac{\hat{F}}{\hat{a}/\mathrm{j}\omega} = \mathrm{j}\omega M.$$

Note that the sign of the imaginary part of the impedance changes if the $\mathrm{e}^{-\mathrm{j}\omega t}$ convention is used instead of the $\mathrm{e}^{\mathrm{j}\omega t}$ convention, because integration with respect to time with the former convention corresponds to multiplication with $-1/\mathrm{j}\omega$. This is a general result – all impedances should be complex conjugated if the convention is changed.

Example 5.2 Mechanical impedance of a spring: It takes a force of $F = \xi \cdot K$ to stretch a spring with the stiffness K by a length of ξ (Hooke's law); and this is also the case if the force is harmonic; therefore the mechanical impedance of the spring is

$$Z_{\mathrm{m}} = \frac{\hat{F}}{\hat{v}} = \frac{\hat{F}}{\mathrm{j}\omega\hat{\xi}} = \frac{K}{\mathrm{j}\omega}.$$

Example 5.3 A mechanical oscillator: A simple mechanical oscillator consists of a mass M suspended from a spring with a stiffness constant of K, as sketched in Figure 5.1. In order to set the mass into vibrations one will have to move the mass *and* displace the spring from its equilibrium value. It follows that the mechanical impedance of this system is the sum of the impedance of the mass and the impedance of the spring,

$$Z_{\mathrm{m}} = \mathrm{j}\omega M + \frac{K}{\mathrm{j}\omega} = \mathrm{j}\left(\omega M - \frac{K}{\omega}\right) = \mathrm{j}\omega M\,(1 - (\omega_0/\omega)^2),$$

where

$$\omega_0 = \sqrt{K/M}$$

is the angular resonance frequency. Note that the impedance is zero at the resonance frequency, indicating that even a very small harmonic force at this frequency will generate an infinite velocity. In practice there will always be some losses, of course, so the impedance is very small but not zero at the resonance frequency.

5.2   Acoustic Impedance

The *acoustic impedance* is associated with average properties on a surface. This quantity is mainly used under conditions where the sound pressure is more or less constant on the surface. It is defined as the complex ratio of the average sound pressure $\hat{p}_{av}$ to the *volume velocity*, which is the surface integral of the normal component of the particle velocity,

$$\hat{q} = \int_S \hat{\mathbf{u}} \cdot d\mathbf{S}, \tag{5.3}$$

where S is the surface area. Thus the acoustic impedance is

$$Z_a = \hat{p}_{av}/\hat{q}. \tag{5.4}$$

The unit is $kg\,m^{-4}s^{-1}$. Since the total force acting on the surface equals the product of the average sound pressure and the area, and since $\hat{q} = S\hat{u}_n$ if the velocity is uniform, it can be seen that there is a simple relation between the two impedance concepts under such conditions:

$$Z_m = Z_a S^2. \tag{5.5}$$

This equation makes it possible to calculate the force it would take to drive a (real or fictive) massless piston with the velocity $\hat{u}_n$. If the piston is real, the impedance is called the *radiation impedance*. This quantity is used for describing the load on, for example, a loudspeaker membrane caused by the medium.[1]

The concept of acoustic impedance is essentially associated with approximate low-frequency models. For example, as we shall see in Section 7.5, it is a very good approximation to assume that the sound field in a tube is one-dimensional when the wavelength is long compared with the cross-sectional dimensions of the tube. Under such conditions the sound field can be described by Equations (3.11) and (3.12), and a tube of a given

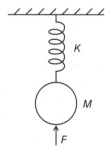

Figure 5.1   A mass hanging from a spring

---

[1] The load of the medium on a vibrating piston can be described either in terms of the acoustic radiation impedance (the ratio of sound pressure to volume velocity) or the mechanical radiation impedance (the ratio of force to velocity).

length behaves as an acoustic two-port.[2] It is possible to calculate the transmission of sound through complicated systems of pipes using fairly simple considerations based on the assumption of continuity of the sound pressure and the volume velocity at each junction; see Section 7.3.[3] The acoustic impedance is also useful in studying the properties of acoustic transducers. Such transducers are usually much smaller than the wavelength in a significant part of the frequency range. This makes it possible to employ so-called lumped parameter models where the system is described by an analogous electrical circuit composed of simple lumped element, inductors, resistors and capacitors, representing masses, losses and springs [1]. Finally it should be mentioned that the acoustic impedance can be used for describing the acoustic properties of materials exposed to normal sound incidence.

Example 5.4 Input impedance of a rigidly terminated tube: The acoustic input impedance of a tube terminated by a rigid cap can be deduced from Equations (3.11) and (3.12) (with $x = -l$),

$$Z_a = -j \frac{\rho c}{S} \cot kl,$$

where l is the length of the tube and S is its cross-sectional area. Note that the impedance goes to infinity when l equals a multiple of half a wavelength, indicating that it would take an infinitely large force to drive a piston at the inlet of the tube at these frequencies (see Figure 5.2). Conversely, the impedance is zero when l equals an odd-numbered multiple of a quarter of a wavelength; at these frequencies the sound pressure on a vibrating piston at the inlet of the tube would vanish.

At low frequencies the acoustic impedance of the rigidly terminated tube analysed in Example 5.4 can be simplified. The factor $\cot kl$ approaches $1/(kl)$, and the acoustic impedance becomes

$$Z_a \simeq -j \frac{\rho c}{Slk} = \frac{\rho c^2}{j \omega V}, \tag{5.6}$$

where $V = Sl$ is the volume of the tube, indicating that the air in the tube acts as a spring; cf. Example 5.2. Thus the acoustic impedance of a cavity much smaller than the wavelength is spring-like, with a stiffness that is inversely proportional to the volume and independent of the shape of the cavity. Since, from Equation (2.10),

$$\rho c^2 = \gamma p_0, \tag{5.7}$$

---

[2] 'Two-port' is a term from electric circuit theory denoting a network with two terminals. Such a network is completely described by the relations between four quantities, the voltage and current at the input terminal and the voltage and current at the output terminal. By analogy, an acoustic two-port is completely described by the relations between the sound pressures and the volume velocities at the two terminals. As shown in Section 7.3 such relations can easily be derived from Equations (3.11) and (3.12) in the case of a cylindrical tube.

[3] Such systems act as acoustic filters. Silencers (or mufflers) are composed of coupled tubes.

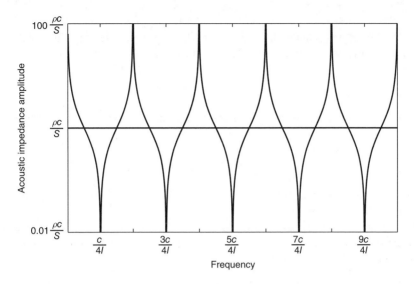

Figure 5.2 The acoustic input impedance of a tube terminated rigidly

it can be seen that the acoustic impedance of a cavity at low frequencies also can be written

$$Z_a = \frac{\gamma p_0}{j\omega V},$$ (5.8)

in agreement with Example 2.1.

Example 5.5 A Helmholtz resonator: A Helmholtz resonator is the acoustic analogue to the simple mechanical oscillator described in Example 5.3; see Figure 5.3. The dimensions of the cavity are much smaller than the wavelength; therefore it behaves as a spring with the acoustic impedance

$$Z_a = \frac{\rho c^2}{j\omega V},$$

where V is the volume; cf. Equation (5.6). The air in the neck moves back and forth uniformly as if it were incompressible; therefore the air in the neck behaves as a lumped mass with the mechanical impedance

$$Z_m = j\omega \rho S l_{eff},$$

where l_{eff} is the effective length and S is the cross-sectional area of the neck. (The effective length of the neck is somewhat longer than the physical length, because some of the air just outside the neck is moving along with the air in the neck.) The corresponding acoustic impedance follows from Equation (5.5):

$$Z_a = \frac{j\omega\rho l_{eff}}{S}.$$

By analogy with Example 5.3 we conclude that the angular resonance frequency is

$$\omega_0 = c\sqrt{\frac{S}{Vl_{eff}}}.$$

Note that the resonance frequency is independent of the static pressure and the density of the medium.

It is intuitively clear that a larger volume or a longer neck would correspond to a lower frequency, but it is perhaps less obvious that a neck with a smaller cross section gives a lower frequency.

5.3 Specific Impedance, Wave Impedance and Characteristic Impedance

In theoretical work the properties of acoustic materials such as for example porous layers are usually described in terms of their *specific acoustic surface impedance*, that is, the ratio of the sound pressure to the resulting normal component of the particle velocity.[4] The unit is $\mathrm{kgm^{-2}s^{-1}}$. If the local ratio of pressure to normal component of particle velocity on the surface is independent of the nature of the sound field (i.e., independent of whether the incident field is a plane wave of a spherical wave), then the material is *locally reacting*.

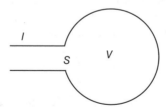

Figure 5.3 A Helmholtz resonator

---

[4] In most practical applications the properties of acoustic materials are described in terms of absorption coefficients (sometimes called absorption factors), assuming either normal, oblique or diffuse sound incidence. The absorption coefficient is defined in terms of a power ratio; see Sections 6.6 and 8.3. It is possible to calculate the absorption coefficient of a material from its specific acoustic impedance, but not the impedance from the absorption coefficient. (An interesting attempt to predict impedances from measured absorption data on the basis of an underlying impedance model has recently been presented, though [2].)

Reflection from an infinite, locally reacting impedance plane

The reflection of sound from an impedance plane depends on the nature of the sound field. A normally incident plane wave is reflected as follows from an infinite, plane, locally reacting surface at $x = 0$ with the specific surface impedance Z_s. The sound pressure can be written as

$$\hat{p} = p_i e^{j(\omega t + kx)} + p_r e^{j(\omega t - kx)} = p_i e^{j\omega t}(e^{jkx} + R e^{-jkx})$$

(cf. Equation (3.15)), and the particle velocity in the x-direction is

$$\hat{u}_x = \frac{p_i e^{j\omega t}}{\rho c}(-e^{jkx} + R e^{-jkx}).$$

From the boundary condition it follows that

$$\frac{\hat{p}(0)}{-\hat{u}_x(0)} = \rho c \frac{1 + R}{1 - R} = Z_s$$

from which an expression for the reflection factor can be derived,

$$R = \frac{\frac{Z_s}{\rho c} - 1}{\frac{Z_s}{\rho c} + 1}.$$

If a similar analysis is carried out for an obliquely incident plane wave (see Figure 5.4) an angle dependent reflection factor is found,

$$R(\theta) = \frac{\frac{Z_s}{\rho c} \cos \theta - 1}{\frac{Z_s}{\rho c} \cos \theta + 1}.$$

The 'spherical reflection factor', which is the exact solution to the reflection problem for spherical sound incidence (generated by a source of the type described by Equation (3.39)), is far more complicated than the plane wave reflection factor; see, e.g., [3].

Example 5.6 Conditions for the interface between fluids being locally reacting: The interface between two fluids studied in Section 3.2 does not in general behave as a locally reacting impedance, because the local ratio of the sound pressure to the normal component of the particle velocity depends on the angle of incidence as can be deduced from Equation (3.33). However, if the speed of sound in fluid no 2 is significantly lower than the speed of sound in fluid no 1, then the wave motion in fluid no 2 will be in the direction normal to the surface irrespective of the sound incidence in fluid no 1; cf. Equation (3.29); and the surface is locally reacting.

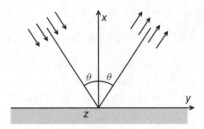

Figure 5.4   An obliquely incident plane wave reflected by a locally reacting surface

The local ratio of the sound pressure to the particle velocity component in the direction of sound propagation in a *sound field* (not on a surface) is occasionally referred to as the *wave impedance*, but often it is called the specific impedance. The unit is $\mathrm{kg\,m^{-2}\,s^{-1}}$.

Yet another impedance concept, the characteristic impedance, has already been introduced. This quantity describes an important quantity of the medium, as we have seen in Section 3.2. As demonstrated in Chapter 3, the complex ratio of the sound pressure to the particle velocity (the wave impedance) in a plane propagating wave equals the characteristic impedance of the medium (cf. Equation (3.10)), and it approximates this value in a free field far from the source (cf. Equation (3.40)).

References

[1] W. Marshall Leach, Jr.: *Introduction to Electroacoustics and Audio Amplifier Design* (3rd edition). Kendall/Hunt Publishing Company, Dubuque, IA (2003).
[2] J.H. Rindel: An impedance model for estimating the complex pressure reflection factor. *Proceedings of Forum Acusticum*, Aalborg, Denmark (2011).
[3] E.M. Salomons: *Computational Atmospheric Acoustics*. Kluwer Academic Publishers (2001).

6

Sound Energy, Sound Power, Sound Intensity and Sound Absorption

6.1 Introduction

The most important acoustic quantity is the sound pressure, which is an acoustic first-order quantity. However, sources of sound emit sound power, and sound fields are also energy fields in which potential and kinetic energies are generated, transmitted and dissipated. In spite of the fact that the radiated sound power is a negligible part of the energy conversion of almost any sound source, energy considerations are of enormous practical importance in acoustics. In 'energy acoustics' sources of noise are described in terms of their sound power, acoustic materials are described in terms of the fraction of the incident sound power that is absorbed, and the sound insulation of partitions is described in terms of the fraction of the incident sound power that is transmitted, the underlying (usually tacit) assumptions being that these properties are independent of the particular circumstances. None of these assumptions is true in the strict sense of the word. However, they are usually good approximations in a significant part of the audible frequency range, and alternative methods based on linear quantities are vastly more complicated than the simple energy balance considerations of energy acoustics.

Sound intensity is a measure of the flow of acoustic energy in a sound field. More precisely, the sound intensity $\mathbf{I}$ is a vector quantity defined as the time average of the net flow of sound energy through a unit area in a direction perpendicular to the area. The dimensions of the sound intensity are energy per unit time per unit area (W/m^2). Although acousticians have attempted to measure this quantity since the 1930s, the first reliable measurements of sound intensity under laboratory conditions did not occur until the middle of the 1970s. Commercial sound intensity measurement systems came on the market in the beginning of the 1980s, and the first international standards for measurements using sound intensity and for instruments for such measurements were issued in the middle of

Fundamentals of General Linear Acoustics, First Edition. Finn Jacobsen and Peter Møller Juhl.
© 2013 John Wiley & Sons, Ltd. Published 2013 by John Wiley & Sons, Ltd.

the 1990s. A description of the history of the development of sound intensity measurement is given in Fahy's monograph *Sound Intensity* [1].

The advent of sound intensity measurement systems in the 1980s had significant influence on noise control engineering. Sound intensity measurements make it possible to determine the sound power of sources without the use of costly special facilities such as anechoic and reverberation rooms, and sound intensity measurements have now routinely been used for decades in the determination of the sound power of machinery and other sources of noise *in situ*. Other important applications of sound intensity include the identification and rank ordering of partial noise sources, visualisation of sound fields, determination of the transmission losses of partitions, and determination of the radiation efficiencies of vibrating surfaces.

The sound intensity method is not without problems, though. Some people consider the method very difficult to use, and it cannot be denied that more knowledge is required in measuring sound intensity than in, say, using an ordinary sound level meter. The difficulties are mainly due to the fact that the accuracy of sound intensity measurements with a given measurement system depends very much on the sound field under study. Another problem is that the distribution of the sound intensity in the near field of a complex source is far more complicated than the distribution of the sound pressure, indicating that sound fields can be much more complicated than realised before the advent of this measurement technique. The problems are reflected in the extensive literature on the errors and limitations of sound intensity measurement and in the fairly complicated international and national standards for sound power determination using sound intensity, ISO 9614-1, ISO 9614-2, ISO 9614-3 and ANSI S12.12.

The purpose of this chapter is to introduce 'energy acoustics' and give a brief but nevertheless relatively detailed overview of sound intensity and its measurement and applications.

6.2 Conservation of Sound Energy

It can be shown [1–3] that the instantaneous potential energy density in a sound field (the potential sound energy per unit volume) is given by the expression

$$w_{\text{pot}}(t) = \frac{p^2(t)}{2\rho c^2}.$$ (6.1)

This quantity describes the energy stored in a small volume of the medium because of compression or rarefaction; the phenomenon is analogous to the potential energy in a compressed or elongated spring.

The instantaneous kinetic energy density in a sound field (the kinetic energy per unit volume) is [1–3]

$$w_{\text{kin}}(t) = \frac{1}{2}\rho u^2(t),$$ (6.2)

where $u(t)$ is the magnitude of the particle velocity vector $\mathbf{u}(t)$. This quantity describes the energy in a small volume of the medium represented by the moving mass of the particles of the medium.

The instantaneous sound intensity is the product of the sound pressure and the particle velocity,

$$\mathbf{I}(t) = p(t)\mathbf{u}(t). \tag{6.3}$$

This quantity, which is a vector, expresses the magnitude and direction of the instantaneous flow of sound energy per unit area, since force per unit area (i.e., pressure) times distance per unit time (i.e., velocity) represents power per area. As shown in the following, one can derive a relation from Equation (6.3) that demonstrates conservation of sound energy. The divergence of the instantaneous sound intensity $\mathbf{I}(t)$ is

$$\nabla \cdot \mathbf{I}(t) = \nabla \cdot (p(t)\mathbf{u}(t)) = p(t)\nabla \cdot \mathbf{u}(t) + \mathbf{u}(t) \cdot \nabla p(t). \tag{6.4}$$

If we combine the two terms on the right-hand side with the linear acoustic equations (Equations (2.12) and (2.13)), we get

$$\nabla \cdot \mathbf{I}(t) = -\frac{p(t)}{\rho c^2}\frac{\partial p(t)}{\partial t} - \rho\mathbf{u}(t) \cdot \frac{\partial \mathbf{u}(t)}{\partial t} = -\frac{\partial}{\partial t}\left(\frac{p^2(t)}{2\rho c^2} + \frac{\rho u^2(t)}{2}\right). \tag{6.5}$$

The quantity in the right-hand parenthesis is recognised as the sum of the instantaneous potential energy density and the instantaneous kinetic energy density, so all in all we can conclude that

$$\nabla \cdot \mathbf{I}(t) = -\frac{\partial w_{\text{tot}}(t)}{\partial t}, \tag{6.6}$$

where $w_{\text{tot}}(t)$ is the total instantaneous energy density.[1] This is the equation of conservation of sound energy, which expresses the simple fact that the rate of decrease of the sound energy density at a given position in a sound field (represented by the right-hand term) is equal to the rate of the flow of sound energy diverging away from the point (represented by the left-hand term).

That Equation (6.6) represents a conservation law is perhaps easier to see from the global version. This is obtained using Gauss's theorem, according to which the local divergence of sound energy integrated over a given volume equals the total net outflow of sound energy through the surface of the volume,

$$\int_V \nabla \cdot \mathbf{I}(t)\mathrm{d}V = \int_S (\mathbf{I}(t) \cdot \mathbf{n})\mathrm{d}S, \tag{6.7}$$

where S is the area of a surface around the source, $\mathbf{n}$ is the normal vector (pointing outwards) and V is the volume contained by the surface. Combining with Equation (6.6) gives

$$\int_S (\mathbf{I}(t) \cdot \mathbf{n})\mathrm{d}S = -\frac{\partial}{\partial t}\left(\int_V w_{\text{tot}}(t)\,\mathrm{d}V\right) = -\frac{\partial E_a}{\partial t}, \tag{6.8}$$

which shows that the total net outflow of sound energy per unit time through the surface equals the (negative) rate of change of the total sound energy within the surface, E_a. In

---

[1] The derivation based on the linear acoustic Equations (2.12) and (2.13) is due to Kirchhoff. Strictly speaking the acoustic energy corollary (Equation (6.5)) should be derived on the basis of the full, non-linear equations rather than the linearised equations, and afterwards reduced to second order. However, since various terms cancel out the result is the same [2–4].

other words, the rate of change of the sound energy within a closed surface is identical with the surface integral of the normal component of the instantaneous sound intensity, $\mathbf{I}(t)$.

In practice we are often concerned with the time-averaged sound intensity in stationary sound fields. For simplicity we shall use the symbol $\mathbf{I}$ for this quantity (rather than $\langle \mathbf{I} \rangle_t$), that is,

$$\mathbf{I} = \langle p(t)\mathbf{u}(t) \rangle_t. \tag{6.9}$$

Examination of Equation (6.6) leads to the conclusion that the divergence of the time averaged sound intensity is identically zero unless we are at a source point,

$$\nabla \cdot \mathbf{I} = 0, \tag{6.10}$$

from which it follows that in a stationary sound field the time-average of the instantaneous net flow of sound energy out of a given closed surface is zero unless there is generation or dissipation of sound power within the surface, that is,

$$\int_S (\mathbf{I}(t) \cdot \mathbf{n}) dS = 0, \tag{6.11}$$

irrespective of the presence of sources *outside* of the surface. If the surface encloses a steady sound source that radiates the sound power P_a then the time-average of the net flow of sound energy out of the surface is equal to the net sound power of the source, that is,

$$\int_S (\mathbf{I}(t) \cdot \mathbf{n}) dS = P_a, \tag{6.12}$$

irrespective of the presence of other steady sources outside the surface and irrespective of the shape of the surface. This important equation is the basis of sound power determination using sound intensity. The usefulness of describing a source in terms of its sound power is due to the fact that this quantity is essentially independent of the surroundings of the source, unlike the sound pressure generated by the source at a given distance.

Example 6.1 Linear superposition of sound pressure and particle velocity: In a sound field generated by two sources we can write

$$p(t) = p_1(t) + p_2(t)$$

and

$$\mathbf{u}(t) = \mathbf{u}_1(t) + \mathbf{u}_2(t)$$

(linear superposition), from which it follows that

$$\mathbf{I} = \langle (p_1(t) + p_2(t))(\mathbf{u}_1(t) + \mathbf{u}_2(t)) \rangle_t = \mathbf{I}_1 + \mathbf{I}_2 + \langle p_1\mathbf{u}_2(t) \rangle_t + \langle p_2\mathbf{u}_1(t) \rangle_t.$$

Note that when the sources are uncorrelated the two sound intensity vectors are simply added, since the time average of each cross term is zero. However, this is not the case when the sources are correlated.

If the sound field is harmonic with angular frequency $\omega = 2\pi f$ we can make use of the usual complex representation of the sound pressure and the particle velocity.

However, second-order quantities require a special formulation; see Appendix A. Expressed in terms of these quantities Equation (6.9) becomes

$$I_r = \frac{1}{2}\mathrm{Re}\{\hat{p}\hat{u}_r^*\} = \frac{|\hat{p}||\hat{u}_r|}{2}\cos\varphi = \frac{|\hat{u}_r|^2}{2}\mathrm{Re}\{Z_s\} = \frac{|\hat{p}|^2}{2}\mathrm{Re}\{Y_s\}, \qquad (6.13)$$

where φ is the phase angle between sound pressure and particle velocity, Z_s is the wave impedance and Y_s is the wave admittance (the reciprocal of the wave impedance). (For simplicity we consider only the component of the particle velocity in the r-direction here.) In a plane progressive wave the sound pressure and the particle velocity are in phase ($\varphi = 0$) and related by the characteristic impedance of the medium, ρc (cf. Equation (3.10)):

$$\hat{p} = \rho c \hat{u}_r. \qquad (6.14)$$

Thus for a plane wave the sound intensity is

$$I_r = \langle p(t)u_r(t)\rangle_t = \frac{\langle p^2(t)\rangle_t}{\rho c} = \frac{1}{2}\frac{|\hat{p}|^2}{\rho c} = \frac{p_{rms}^2}{\rho c}. \qquad (6.15)$$

In this case the sound intensity is simply related to the mean square sound pressure p_{rms}^2, which can be measured with a single microphone.[2] Equation (6.15) is also valid in the simple spherical sound field generated by a monopole in free space, irrespective of the distance to the source. However, in the general case the sound intensity is *not* simply related to the sound pressure, and both the sound pressure and the particle velocity must be measured simultaneously and their instantaneous product time-averaged. This requires the use of a more complicated device than a single microphone.

Example 6.2 Divergence of time-averaged intensity: The divergence of the time averaged sound intensity can be written

$$\nabla \cdot \mathbf{I} = \frac{1}{4}\nabla \cdot (\hat{p}\hat{\mathbf{u}}^* + \hat{p}^*\hat{\mathbf{u}}) = \frac{1}{4}(\hat{p}\nabla \cdot \hat{\mathbf{u}}^* + \hat{p}^*\nabla \cdot \hat{\mathbf{u}} + \hat{\mathbf{u}}^* \cdot \nabla\hat{p} + \hat{\mathbf{u}} \cdot \nabla\hat{p}^*).$$

Expressed in complex notation Equations (2.12) and (2.13) become

$$\nabla \cdot \hat{\mathbf{u}} = -\frac{j\omega}{\rho c^2}\hat{p}$$

and

$$\nabla\hat{p} = -j\omega\rho\hat{\mathbf{u}}.$$

When these two equations are inserted into the expression for the divergence of **I** Equation (6.10) results.

---

[2] Equation (6.15) implies that $L_I \simeq L_p$ under normal ambient conditions, since

$$L_I = 10\log\left(\frac{I}{I_{ref}}\right) = 10\log\left(\frac{p_{rms}^2}{p_{ref}^2}\right) + 10\log\left(\frac{p_{ref}^2}{\rho c I_{ref}}\right) = L_p + 10\log\left(\frac{p_{ref}^2}{\rho c I_{ref}}\right)$$

where the last term is small, about -0.1 dB, at 101.3 kPa and 20°.

Example 6.3 Sound intensity in a standing wave: The sound intensity in the plane standing wave analysed in Section 3.1 is, from Equations (3.13) and (3.14),

$$I_x = \frac{1}{2}\mathrm{Re}\{2p_+ \cos(kx)jp_+^* \sin(kx)\} = 0.$$

The sound pressure and the particle velocity are in quadrature and thus the time-averaged sound intensity is zero, in agreement with the fact that the rigid plane at $x = 0$ reflects all sound

Example 6.4 Sound energy transmission between fluids at normal incidence: We can now supplement the analysis of the transmission of sound between two fluids in Section 3.2 by looking at the transmission of sound energy. From Equations (3.25) and (6.15) we conclude that for normal incidence the ratio of transmitted to incident sound intensity is

$$\frac{I_t}{I_i} = \frac{p_t^2}{p_i^2}\frac{\rho_1 c_1}{\rho_2 c_2} = \frac{4\rho_1 c_1 \rho_2 c_2}{(\rho_1 c_1 + \rho_2 c_2)^2},$$

which is seen to be independent of whether sound is transmitted from fluid no 1 to fluid no 2, unlike the ratio of the amplitude of the transmitted sound pressure to the amplitude of the incident sound pressure.

Example 6.5 Sound energy transmission between fluids at oblique incidence: In Section 3.2 we analysed transmission of sound between fluids in case of an obliquely incident plane wave from fluid no 1 striking the interface to fluid no 2, and observed that when the speed of sound in fluid no 2 exceeds the speed of sound in fluid no 1 and the angle to the normal of the incident wave, θ, exceeds a certain 'critical angle', then the sound pressure in fluid no 2 decays exponentially with the distance to the surface. From Equation (3.31),

$$\hat{p}_2 = p_t e^{j\omega t} e^{-\gamma x} e^{-jk_2 y \sin\theta_2},$$

and Euler's equation of motion we can calculate the particle velocity in the normal direction in this evanescent wave,

$$\hat{u}_2 = -\frac{1}{j\omega\rho_2}\frac{\partial \hat{p}_2}{\partial x} = -j\frac{p_t}{\rho_2 c_2}\frac{\alpha}{k_2}e^{j\omega t}e^{-\gamma x}e^{-jk_2 y \sin\theta_2}.$$

It is apparent that the sound pressure and the particle velocity are in quadrature, and thus the corresponding sound intensity is zero (cf. Example 6.3); there is no flow of sound energy away from the surface in fluid no 2.

Example 6.6 Sound intensity in a simple spherical field: In a simple spherical sound field we have the following relation between the sound pressure and the particle velocity (Equation (3.40)),

$$\hat{u}_r = \frac{\hat{p}}{\rho c}\left(1 + \frac{1}{jkr}\right).$$

It is apparent that the component of the particle velocity in phase with the sound pressure is $\hat{p}/\rho c$, just as in a plane propagating wave, which explains why the sound intensity equals $|\hat{p}|^2/2\rho c$.

Example 6.7 The free-field method of measuring sound power: The free-field method of estimating the sound power of a source relies on the fact that the plane wave expression for the sound intensity, Equation (6.15), is a good approximation sufficiently far from any finite source in free space (the sound field becomes 'locally plane', in agreement with what we shall hear about in Chapter 9, the Sommerfeld radiation condition). In practice an anechoic room or free space out of doors is required.

6.3 Active and Reactive Intensity

For later reference we will derive a relation between the sound intensity and the gradient of the phase of the sound pressure in a harmonic sound field. If we write the complex sound pressure in the form of an amplitude and a phase,

$$\hat{p} = |\hat{p}|e^{j\phi}, \tag{6.16}$$

and make use of Euler's equation of motion, Equation (2.13), the expression for the sound intensity becomes

$$\mathbf{I} = \frac{1}{2}\mathrm{Re}\{\hat{p}\hat{u}^*\} = \frac{1}{2}\mathrm{Re}\left\{\hat{p}\left(\frac{-1}{j\omega\rho}\nabla\hat{p}\right)^*\right\} = \frac{1}{2}\mathrm{Re}\left\{|\hat{p}|e^{j\phi}\left(\frac{-1}{j\omega\rho}\nabla\left(|\hat{p}|e^{j\phi}\right)\right)^*\right\}$$

$$= \frac{1}{2}\mathrm{Re}\left\{\frac{|\hat{p}|e^{j\phi}}{j\omega\rho}\left(e^{-j\phi}\nabla|\hat{p}| - je^{-j\phi}|\hat{p}|\nabla\phi\right)\right\}$$

$$= -\frac{|\hat{p}|^2}{2\omega\rho}\nabla\phi = -\frac{|\hat{p}|^2}{2\rho c}\frac{\nabla\phi}{k}, \tag{6.17}$$

which shows that the sound intensity equals the product of $p_{\mathrm{rms}}^2/\rho c$ and the gradient of the phase of the sound pressure normalised with the wavenumber. Inspection of this equation leads to the interesting conclusion that the time-averaged sound intensity is orthogonal to the wavefronts, that is, surfaces of constant phase in the sound field [5].

Example 6.8 Pressure phase gradient in a simple spherical sound field: The sound pressure at a distance r from a spherically symmetrical source with is given by Equation (3.39),

$$\hat{p} = \frac{A e^{j(\omega t - kr)}}{r}.$$

Note that the gradient of the phase in the r-direction equals $-k$, just as in a plane propagating sound wave. Inserting into Equation (6.17) shows that

$$I_r = \frac{|\hat{p}|^2}{2\rho c};$$

cf. Example 6.6.

In spite of the diversity of sound fields encountered in practice, some typical sound field characteristics can be identified. For example, the sound field far from the source that generates it has certain well-known properties under free-field conditions, the sound field near a source has other characteristics, and some characteristics are typical of a reverberant sound field, and so on.

We have seen that the sound pressure and the particle velocity are in phase in a plane propagating wave. This is also the case in a free field, sufficiently far from the source that generates the field. Conversely, one of the characteristics of the sound field near a source is that the sound pressure and the particle velocity are partly out of phase (in quadrature). To describe such phenomena one may introduce the concept of *active* and *reactive* sound fields.

It takes four second-order quantities to describe the distributions and fluxes of sound energy in a sound field completely [5–7]: potential energy density, kinetic energy density, active intensity (which is the quantity we usually simply refer to as the intensity), and reactive intensity. The last mentioned of these quantities represents the non-propagating, oscillatory sound energy flux that is characteristic of a sound field in which the sound pressure and the particle velocity are in quadrature, as for instance in the near field of a small source. The reactive intensity is a vector defined as the imaginary part of the product of the complex pressure and the complex conjugate of the particle velocity,

$$\mathbf{J} = \frac{1}{2} \text{Im}\{\hat{p}\hat{\mathbf{u}}^*\} \tag{6.18}$$

(cf. Equation (6.13)). More general time-domain formulations based on the Hilbert transform are also available [7]. Unlike the usual active intensity, the reactive intensity remains a somewhat controversial issue although the quantity was introduced in the middle of the previous century [8], perhaps because the vector $\mathbf{J}$ has no obvious physical meaning [9], or perhaps because describing an oscillatory flux by a time-averaged vector seems peculiar to some. However, even though the reactive intensity is of no obvious direct practical use it nevertheless is quite convenient that we have a quantity that makes it possible to describe and quantify the particular sound field conditions in the near field of sources in a precise manner.

Very near a sound source the reactive field is usually stronger than the active field. However, the reactive field dies out rapidly with increasing distance to the source. Therefore, even at a fairly moderate distance from the source, the sound field is dominated by the active field. The extent of the reactive field depends on the frequency, and the dimensions and the radiation characteristics of the sound source. In most practical cases the reactive field may be assumed to be negligible at a distance greater than, say, half a metre from the source.

Example 6.9 Wave impedance near a monopole: From Example 6.6 we can calculate the wave impedance at a distance r from a monopole

$$Z_s = \frac{\hat{p}}{\hat{u}_r} = \frac{\rho c}{1 + 1/jkr}.$$

Apparently, this quantity is almost purely imaginary when $kr \ll 1$ indicating that the sound pressure and the particle velocity are nearly 90° out of phase in the near field of the source. Note that the wave impedance is mass-like under such condition, that is, proportional to $j\omega$; as we shall see in Chapter 9 the radiation impedance of a monopole is dominated by the mass term.

Figures 6.1, 6.2 and 6.3 demonstrate the physical significance of the active and reactive intensities. Figure 6.1 shows the result of a measurement at a position about 30 cm (about one wavelength) from a small monopole source, a loudspeaker driven with a band of one-third octave noise. The sound pressure and the particle velocity (multiplied by ρc) are almost identical; therefore the instantaneous intensity is always positive: this is an *active* sound field. Figure 6.2 shows the result of a similar measurement very near the loudspeaker (less than one tenth of a wavelength from the cone). In this case the sound pressure and the particle velocity are almost in quadrature, and as a result the instantaneous intensity fluctuates about zero, that is, sound energy flows back and forth. This is an example of a strongly *reactive* sound field. Finally, Figure 6.3 shows the result of a measurement in a reverberant room several metres from the loudspeaker generating the sound field. Here the sound pressure and the particle velocity appear to be uncorrelated signals; this is neither an active nor a reactive sound field; this is a *diffuse* sound field.

If we combine Equations (6.16) and (6.18) we can derive a relation between the reactive intensity and the gradient of the amplitude of the sound pressure, analogous to Equation (6.17):

$$\mathbf{J} = \frac{1}{2}\text{Im}\{\hat{p}\hat{\mathbf{u}}^*\} = \frac{1}{2}\text{Im}\left\{ \frac{|\hat{p}|e^{j\phi}}{j\omega\rho} \left(e^{-j\phi}\nabla|\hat{p}| - je^{-j\phi}|\hat{p}|\nabla\phi\right)\right\}$$

$$= -\frac{|\hat{p}|\nabla|\hat{p}|}{2\omega\rho} = -\frac{\nabla|\hat{p}|^2}{4\rho ck}, \tag{6.19}$$

This equation shows that the reactive intensity is orthogonal to surfaces of equal sound pressure amplitude ('iso pressure contours') [5].

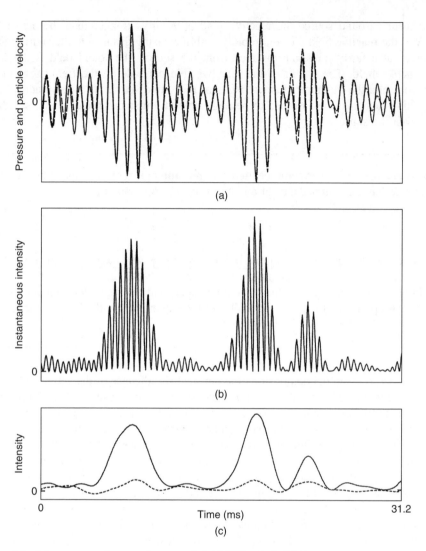

Figure 6.1 Measurement in an active sound field [10]. (a) Solid line, Instantaneous sound pressure; dashed line, instantaneous particle velocity multiplied by the characteristic impedance of air. (b) Instantaneous sound intensity. (c) Solid line, real part of 'complex instantaneous intensity'; dashed line, imaginary part of 'complex instantaneous intensity'. One-third octave noise with a centre frequency of 1 kHz. Reprinted by permission of Elsevier Publishing

The fact that $\mathbf{I}$ is the real part and J is the imaginary part of $1/2 p u^*$ has led to the concept of complex sound intensity,

$$\mathbf{I} + j\mathbf{J} = \frac{1}{2}\hat{p}\hat{\mathbf{u}}^*. \tag{6.20}$$

Note the interesting relation

$$|\mathbf{I}|^2 + |\mathbf{J}|^2 = \frac{|\hat{p}|^2|\hat{\mathbf{u}}|^2}{4}. \tag{6.21}$$

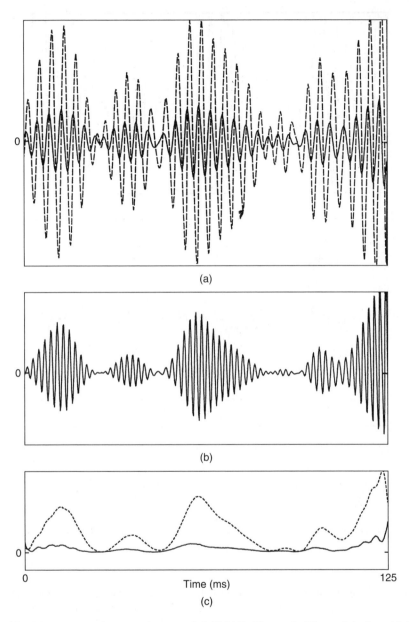

(a)

(b)

(c)

0   Time (ms)   125

Figure 6.2 Measurement in a reactive sound field [10]. Key as in Figure 6.1. One-third octave noise with a centre frequency of 250 Hz. Reprinted by permission of Elsevier Publishing

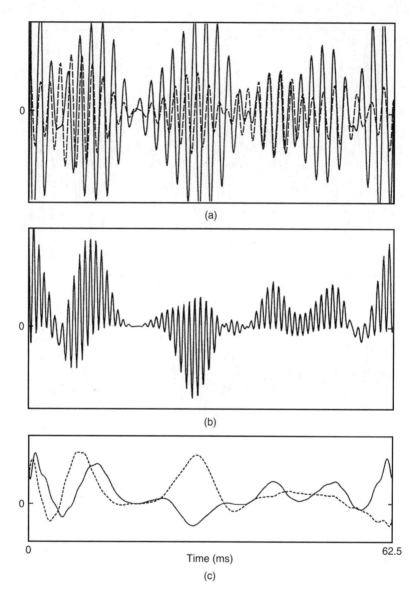

Figure 6.3 Measurement in a diffuse sound field [10]. Key as in Figure 6.1. One-third octave noise with a centre frequency of 500 Hz. Reprinted by permission of Elsevier Publishing

A 'complex instantaneous sound intensity' has also been suggested. As can be seen from Figures 6.1(c), 6.2(c) and 6.3(c) the real and imaginary parts of this quantity represent the envelopes of the active and reactive instantaneous intensity. See [10] for an overview of the various 'sound intensities'.

6.4 Measurement of Sound Intensity

It is far more difficult to measure sound intensity than to measure sound pressure. One problem is that the accuracy depends strongly on the sound field under study; under certain conditions even minute imperfections of the measuring equipment will have a significant influence on the results. With hindsight the 50-year delay from Olson when submitted his application for a patent for an intensity meter in 1931 to when commercial measurement systems came on the market in the beginning of the 1980s is therefore not surprising. See [1].

Many attempts to develop sound intensity probes based on the combination of a pressure transducer and a particle velocity transducer (often referred to as 'pressure-velocity probes' have been described in the literature. For example, a device that combined a pressure microphone with a transducer based on the convection of an ultrasonic beam by the particle velocity 'flow' was produced by Norwegian Electronics for some years. However, the production of this device was stopped in the early 1990s. More recently a micro-machined transducer based on hot wire technology, 'Microflown', has become available for measurement of the particle velocity. A sound intensity probe based on this transducer combined with a small pressure microphone is in commercial production [11, 12], and results demonstrate that it has potential not only for measuring sound intensity [13] but also for a number of other applications [14, 15]. However, one problem with *any* particle velocity transducer, irrespective of the measurement principle, is the influence of airflow. Another serious and only partly solved problem is how to determine the phase correction that is needed when two fundamentally different transducers are combined. Several possible methods of calibrating pressure-velocity probes are described and examined in [16].

All sound intensity measurement systems in commercial production today except the Microflown are based on the 'two-microphone' (or 'p-p') principle, which makes use of two closely spaced pressure microphones and relies on a finite difference approximation to the sound pressure gradient, and the IEC 61043 standard on instruments for the measurement of sound intensity deals exclusively with the two-microphone measurement principle. Accordingly, all the considerations in this chapter concern this measurement principle.

The two-microphone measurement principle employs two closely spaced pressure microphones. The particle velocity is obtained through an approximation to Euler's relation, Equation (2.13), as

$$\tilde{u}_r(t) = -\int_{-\infty}^{t} \frac{p_2(\tau) - p_1(\tau)}{\rho \Delta r} d\tau, \qquad (6.22)$$

where p_1 and p_2 are the sound pressure signals from the two microphones, Δr is the microphone separation distance, and τ is a dummy time variable. The tilde indicates that this is an estimate, in this case based on a finite difference approximation to the real sound pressure gradient. The sound pressure at the centre of the probe is estimated as

$$\tilde{p} = \frac{p_1 + p_2}{2}, \qquad (6.23)$$

and the time-averaged intensity component in the axial direction is, from Equation (6.9),

$$\tilde{I}_r = \langle \tilde{p}(t) \tilde{u}_r(t) \rangle_t = \frac{1}{2\rho \Delta r} \left\langle (p_1(t) + p_2(t)) \int_{-\infty}^{t} (p_1(\tau) + p_2(\tau)) \mathrm{d}\tau \right\rangle_t. \tag{6.24}$$

Some sound intensity analysers use Equation (6.24) to measure the intensity in frequency bands (one-third octave bands, for example). Another type calculates the intensity from the imaginary part of the cross spectrum of the two microphone signals, S_{12},

$$\tilde{I}_r(\omega) = -\frac{1}{\omega \rho \Delta r} \mathrm{Im}\{S_{12}(\omega)\}. \tag{6.25}$$

The time domain formulation is equivalent to the frequency domain formulation, and in principle Equation (6.25) gives exactly the same result as Equation (6.24) when the intensity spectrum is integrated over the frequency band of concern.[3] The frequency domain formulation makes it possible to determine sound intensity with an FFT analyser.

The most common microphone arrangement is known as 'face-to-face'. This arrangement, with a solid spacer between the microphones, has the advantage that the 'acoustic distance' is well defined [17]. A face-to-face sound intensity probe produced by Brüel and Kjær is shown in Figure 6.4.

There are many sources of error in the measurement of sound intensity, and a considerable part of the sound intensity literature has been concerned with identifying and studying such errors. Some of the sources of error are fundamental and others are associated with various technical deficiencies. As mentioned, one complication is that the accuracy depends very much on the sound field under study; under certain conditions even minute imperfections in the measuring equipment will have a significant influence. Another disturbing complication is that small local errors are sometimes amplified into large global errors when the intensity is integrated over a closed surface, as pointed out by Pope [18].

The following is an overview of some of the sources of error in the measurement of sound intensity. Those who make sound intensity measurements with an intensity probe based on the two-microphone measurement principle should know about the limitations imposed by

- the finite difference error [19],
- errors due to scattering and diffraction [17, 20, 21], and
- instrumentation phase mismatch [22, 23].

In what follows these sources of error will be described.

---

[3] This follows from the fact that Equation (6.24) expressed in the frequency domain has the form

$$\tilde{I}_r = \frac{1}{2\pi} \int_{-\infty}^{\infty} S_{\tilde{p}\tilde{u}_r}(\omega)\mathrm{d}\omega = \frac{1}{4\pi \rho \Delta r} \int_{-\infty}^{\infty} \left(\frac{S_{11}(\omega) - S_{22}(\omega) + S_{12}^*(\omega) - S_{12}(\omega)}{j\omega} \right) \mathrm{d}\omega$$

$$= -\frac{1}{2\pi \rho \Delta r} \int_{-\infty}^{\infty} \frac{S_{12}(\omega)}{j\omega} \mathrm{d}\omega = -\frac{1}{2\pi \rho \Delta r} \int_{-\infty}^{\infty} \frac{\mathrm{Im}\{S_{12}(\omega)\}}{\omega} \mathrm{d}\omega,$$

where the last equation sign follows from the fact that only the imaginary part of the cross spectrum (which is an odd function of the frequency) contributes to the integral. See Appendix B.

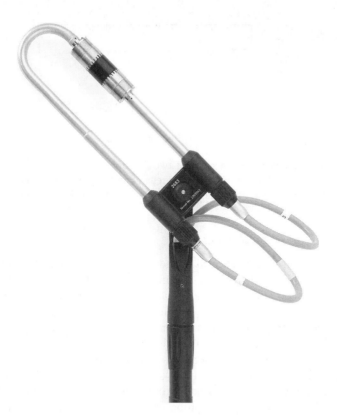

Figure 6.4   Sound intensity probe manufactured by Brüel and Kjær. The microphones are mounted 'face-to-face' with a 'spacer' between them (copyright © Brüel and Kjær)

6.4.1   Errors Due to the Finite Difference Approximation

The most fundamental limitation of the two-microphone measurement principle is due to the fact that the sound pressure gradient is approximated by a finite difference of pressures at two discrete points. This obviously imposes an upper frequency limit that is inversely proportional to the distance between the microphones. The resulting bias error depends on the sound field in a complicated manner [19]. In a plane sound wave of axial incidence the finite difference error, that is, the ratio of the measured intensity $\tilde{I}_r$ to the true intensity I_r, can be shown to be

$$\frac{\tilde{I}_r}{I_r} = \frac{\sin k\,\Delta r}{k\,\Delta r}. \tag{6.26}$$

This relation is shown in Figure 6.5 for different values of the microphone separation distance. The upper frequency limit of intensity probes has generally been considered to be the frequency at which this error is acceptably small. With 12 mm between the microphones (a typical value) this gives an upper limiting frequency of about 5 kHz.

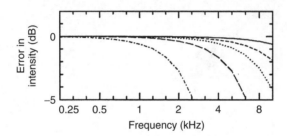

Figure 6.5 Finite difference error of an ideal two-microphone sound intensity probe in a plane wave of axial incidence for different values of the separation distance [20]. Solid line, 5 mm; dashed line, 8.5 mm; dotted line, 12 mm; long dashes, 20 mm; dash-dotted line, 50 mm

6.4.2 Errors Due to Scattering

Equation (6.26) is correct for an ideal sound intensity probe that does not in any way disturb the sound field. In other words, the interference of the microphones on the sound field has been ignored. This would be a good approximation if the microphones were small compared with the distance between them, but it is *not* a good approximation for a typical sound intensity probe such as the one shown in Figure 6.4. The high frequency performance of a real, physical probe is obviously a combination of the finite difference error and the effect of the probe itself on the sound field. In the particular case of the face-to-face configuration it turns out that the two effects to some extent cancel each other for a certain geometry; a numerical and experimental study has shown that the upper frequency limit of such an intensity probe can be extended to about an octave above the limit determined by the finite difference error if the length of the spacer between the microphones equals the diameter. The physical explanation is that the resonance of the cavities in front of the microphones gives rise to a pressure increase that to some extent compensates for the finite difference error. Thus the resulting upper frequency limit of a sound intensity probe composed of half-inch microphones separated by a 12-mm spacer is 10 kHz, which is an octave above the limit determined by the finite difference error when the interference of the microphones on the sound field is ignored [17, 20]; compare Figures 6.5 and 6.6.

6.4.3 Errors Due to Phase Mismatch

Phase mismatch between the two measurement channels is the most serious source of error in measurement of sound intensity, even with the best equipment that is available today. It can be shown that the estimated intensity, subject to a phase error φ_e, to a very good approximation can be written as

$$\tilde{I}_r = I_r - \frac{\varphi_e}{k \Delta r} \frac{p_{rms}^2}{\rho c}, \tag{6.27}$$

that is, the phase error causes a bias error in the measured intensity that is proportional to the phase error and to the mean square pressure [21]. Equation (6.27) is a consequence of Equation (6.17) and can be derived by inserting the actual phase angle between the sound

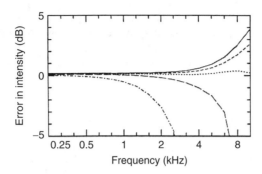

Figure 6.6 Error of a sound intensity probe with half-inch microphones in the face-to-face configuration in a plane wave of axial incidence for different spacer lengths [20]. Solid line, 5 mm; dashed line, 8.5 mm; dotted line, 12 mm; long dashes, 20 mm; dash-dotted line, 50 mm

pressure signals in the sound field plus the phase error due to the measurement system into this expression. Ideally the phase error should be zero, of course. In practice one must, even with state-of-the art equipment, allow for phase errors ranging from about 0.05° at 100 Hz to 2° at 10 kHz. Both the IEC standard and the North American ANSI standard on instruments for the measurement of sound intensity specify performance evaluation tests that ensure that the phase error is within certain limits.

Example 6.10 Pressure phase gradient in a plane wave: In a plane wave and in a simple spherical wave the gradient of the phase of the sound pressure in the direction of propagation equals $-k$ (cf. Example 6.8). It follows that the physical phase difference between the pressures at two points a distance of Δr apart is $k\Delta r$. With 12 mm between the microphones this amounts to about 3° at 250 Hz. Obviously, the phase error introduced by the measurement system should be much smaller than that, say, no larger than 0.3°. Moreover, as shown in Example 6.11 the requirements are much stronger in an interference field.

Example 6.11 Pressure phase gradient in a general sound field: Equation (6.17) shows that the gradient of the phase of the sound pressure can be expressed in terms of the ratio of the mean square pressure to the sound intensity,

$$\frac{\partial \phi}{\partial r} = -k\frac{I_r \rho c}{p_{\text{rms}}^2}.$$

However, since the ratio $(p_{\text{rms}}^2/I_r\rho c)$ may take values of up to, say, ten, under realistic measurement conditions it follows that the phase gradient can easily be ten times smaller than in a plane propagating wave. In other words, it is quite reasonable to require that the phase error of an intensity measurement system at 250 Hz should be *much smaller* than 0.3°, that is, very small indeed; cf. Example 6.10.

Equation (6.27) is often written in the form

$$\tilde{I}_r = I_r + I_o \left(\frac{p_{\text{rms}}^2}{p_o^2} \right) = I_r \left(1 + \frac{I_o \rho c}{p_o^2} \frac{p_{\text{rms}}^2}{I_r \rho c} \right). \tag{6.28}$$

where the *residual intensity* I_o and the corresponding sound pressure p_o,

$$\frac{I_o}{p_o^2/\rho c} = -\frac{\varphi_e}{k \Delta r}, \tag{6.29}$$

have been introduced. The residual intensity is the 'false' sound intensity indicated by the instrument when the two microphones are exposed to the same pressure p_o, for instance in a small cavity. Under such conditions the true intensity is zero, and the indicated intensity I_o should obviously be as small as possible. The right-hand side of Equation (6.28) clearly shows how the error caused by phase mismatch depends on the ratio of the mean square pressure to the intensity in the sound field – in other words on the sound field condition.

Phase mismatch is usually described in terms of the so-called residual pressure-intensity index,

$$\delta_{pI_o} = 10 \log \left(\frac{p_o^2/\rho c}{|I_o|} \right), \tag{6.30}$$

which is just a convenient way of measuring, and describing, the phase error φ_e. With a microphone separation distance of 12 mm the typical phase error mentioned above corresponds to a pressure-residual intensity index of 18 dB in most of the frequency range. The error due to phase mismatch is small provided that

$$\delta_{pI} \ll \delta_{pI_o}, \tag{6.31}$$

where

$$\delta_{pI} = 10 \log \left(\frac{p_{\text{rms}}^2/\rho c}{I_r} \right) \tag{6.32}$$

is the pressure-intensity index of the measurement. The inequality (6.31) is simply a convenient way of expressing that the phase error φ_e of the equipment should be much smaller than the phase angle between the two sound pressure signals in the sound field. A more specific requirement can be expressed in the form

$$\delta_{pI} < L_d = \delta_{pI_o} - K, \tag{6.33}$$

where the quantity

$$L_d = \delta_{pI_o} - K, \tag{6.34}$$

is called 'the dynamic capability' of the instrument and K is 'the bias error factor'. As can be seen from the inequality (6.33) the dynamic capability indicates the maximum acceptable value of the pressure-intensity index of the measurement for a given grade of accuracy. The larger the value of K the smaller is the dynamic capability, the stronger and more restrictive is the requirement, and the smaller is the error. From Equations (3.30) and (3.32) it follows that the inequality (6.33) is equivalent to the requirement

$$\frac{I_o}{p_o^2/\rho c} < \frac{I_r}{p^2/\rho c} 10^{-K/10}, \tag{6.35}$$

which corresponds to requiring that the phase error φ_e should be $10^{K/10}$ times smaller than the phase angle in the sound field. Combined with Equation (6.28) this leads to the conclusion that the condition expressed by the inequality (6.34) and a bias error factor of 7 dB guarantee that the error due to phase mismatch is less than 1 dB; with $K = 10$ dB the error will be less than 0.5 dB. These requirements correspond to the phase error φ_e being five and ten times less than the actual phase angle in the sound field respectively.

Most engineering applications of sound intensity measurements involve integrating the normal component of the intensity over a surface. Integrating both sides of Equation (6.28) over a measurement surface S gives the expression

$$\tilde{P}_a = P_a + \frac{I_o\rho c}{p_o^2}\int_S (p_{\text{rms}}^2/\rho c)\mathrm{d}S = P_a \left(1 + \frac{I_o\rho c}{p_o^2}\frac{\int_S \left(p_{\text{rms}}^2/\rho c\right)\mathrm{d}S}{\int_S (\mathbf{I}\cdot\mathbf{n})\mathrm{d}S}\right), \tag{6.36}$$

which shows that the global version of the inequality (6.33) can be written

$$\Delta_{pI} < \delta_{pI_o} - K, \tag{6.37}$$

where

$$\Delta_{pI} = 10\log\left(\frac{\int_S \left(p_{\text{rms}}^2/\rho c\right)\mathrm{d}S}{\int_S (\mathbf{I}\cdot\mathbf{n})\mathrm{d}S}\right) \tag{6.38}$$

is the global pressure-intensity index of the measurement. Comparing Equations (6.33) and (6.37) shows that this quantity plays the same role in sound power estimation as the pressure-intensity index does in measurements at discrete points. Figure 6.7 shows examples of the global index measured under various conditions. It is obvious that the presence of noise sources outside the measurement surface increases the mean square pressure on the surface, and thus the influence of a given phase error; therefore a phase error, no matter how small, limits the range of measurement.

In practice one should examine whether the inequality (6.37) is satisfied or not whenever there is significant noise from extraneous sources. If the inequality is not satisfied it can be recommended to use a measurement surface somewhat closer to the source than advisable in more favourable circumstances. It may also be necessary to modify the measurement conditions – to shield the measurement surface from strong extraneous sources, for example, or to increase the sound absorption in the room (see Section 8.3.1).

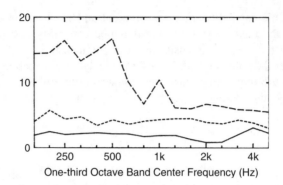

Figure 6.7 The global pressure-intensity index determined under three different conditions. Solid line, Measurement using a 'reasonable' surface; dashed line, measurement using an eccentric surface; long dashes, measurement with strong background noise [24], reprinted by permission of the Institute of Noise Control Engineering

All modern sound intensity analysers can determine the pressure-intensity index concurrently with the actual measurement, so one can easily check whether phase mismatch is a problem or not. Some instruments automatically examine whether the condition (6.37) (or (6.33) in a point measurement) is satisfied or not and give warnings when this is not the case.

By now it should be apparent that even a sound intensity probe of the highest quality will give erroneous results under sufficiently difficult sound field conditions. A standardised verification procedure therefore prescribes that the intensity probe should be exposed to the sound field in a standing wave tube with a specified standing wave ratio; when the sound intensity probe is drawn through this interference field the sound intensity indicated by the measurement system should be within a certain tolerance.

Figure 6.8(a) illustrates how the sound pressure, the particle velocity and the sound intensity vary with position in a one-dimensional interference field with a standing wave ratio of 24 dB. It is apparent that the pressure-intensity index varies strongly with the position in such a sound field. Accordingly, the influence of a given phase error depends on the position. Figure 6.8(b) shows how the sound intensity measured with a certain instrument will deviate from the true value as a function of the position in a standing wave tube with a standing wave ratio of 24 dB, which is the sound field specified in the IEC standard on sound intensity measurement systems. According to this IEC standard deviations within an interval of ± 1.5 dB are acceptable for 'class 1 instruments'.

6.5   Applications of Sound Intensity

Some of the most common practical applications of sound intensity measurements are now discussed briefly.

6.5.1   Sound Power Determination

One of the most important applications of sound intensity measurements is the determination of the sound power of operating machinery *in situ*. Sound power determination using

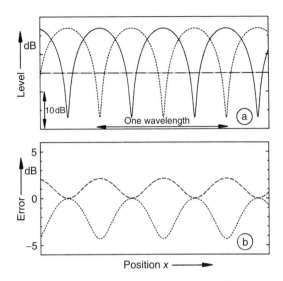

Figure 6.8 (a) Sound pressure level (solid line), particle velocity level (dashed line) and sound intensity level (dash-dotted line) in a standing wave field with a standing wave ratio of 24 dB. (b) Estimation error of a sound intensity measurement system with a residual pressure-intensity index of 14 dB (positive and negative phase error) [25], reprinted by permission of the European Acoustics Association

intensity measurements is based on Equation (6.12), which shows that the sound power of a source is given by the integral of the normal component of the intensity over a surface that encloses the source, also in the presence of other sources outside the measurement surface. Neither an anechoic nor a reverberation room is required. The analysis of errors and limitations presented in Section 6.4 leads to the conclusion that the sound intensity method is suitable

- for stationary sources in stationary background noise provided that $\Delta_{pI} < \delta_{pI_0} - K$.

The method is *not* suitable

- for sources that operate in long cycles (because the sound field will change during the measurement)
- in non-stationary background noise (for the same reason)
- for weak sources of low frequency noise (because of large random errors caused by electrical noise in the microphone signals; see [26]).

The surface integral can be approximated either by sampling at discrete points or by scanning manually or with a robot over the surface. With the scanning approach, the intensity probe is moved continuously over the measurement surface in such a way that the axis of the probe is always perpendicular to the measurement surface. The scanning procedure, which was introduced in the late 1970s on a purely empirical basis, was regarded with much scepticism for more than a decade [27], but has now for many years been generally considered to be more accurate and far more convenient than the procedure based on fixed points [28]. A moderate scanning rate, say $0.5 \, \text{ms}^{-1}$, and a 'reasonable' scan line density should be used, say 5 cm between adjacent lines if the surface is very close to the source, 20 cm if it is further away. One cannot use the scanning method if the source is operating in cycles, though; both the source under test and possible extraneous noise sources must be perfectly stationary.

Usually the measurement surface is divided into a number of segments, each of which will be convenient to scan. One will often determine the pressure-intensity index of each segment, and the accuracy of each partial sound power estimate will of course depend on whether the inequality (6.37) is satisfied or not, but it follows from Equation (6.36) that it is the global pressure-intensity index associated with the entire measurement surface that determines the accuracy of the estimate of the (total) radiated sound power. It may be impossible to satisfy (6.37) on a certain segment, for example because the net sound power passing through the segment takes a very small value because of noise from sources outside the measurement surface, but if the global criterion is satisfied then the total sound power estimate will nevertheless be accurate.

Theoretical considerations seem to indicate the existence of an optimum measurement surface that minimises measurement errors. In practice one uses a surface of a simple shape at some distance, say $25-50 \, \text{cm}$, from the source. If there is a strong reverberant field or significant ambient noise from other sources, the measurement surface should be chosen to be somewhat closer to the source under study.

6.5.2 Noise Source Identification and Visualisation of Sound Fields

A noise reduction project usually starts with the identification and ranking of noise sources and transmission paths, and sound intensity measurements make it possible to determine the partial sound power contribution of the various components directly. Two-dimensional contour plots of the sound intensity normal to a measurement surface can be used in locating noise sources. Visualisation of sound fields, helped by modern computer graphics, contributes to our understanding of radiation and propagation of sound and of diffraction and interference effects. However, sound intensity measurements at many discrete points are very time-consuming, and today one would probably use alternative, more efficient methods for visualising sound fields, say, near field acoustic holography or beamforming, for this task.

6.5.3 Transmission Loss of Structures and Partitions

The conventional measure of the sound insulation of panels and partitions is the transmission loss (also called sound reduction index), which is the ratio of incident to transmitted sound power in logarithmic form. The traditional method of measuring this quantity requires a transmission suite consisting of two vibration-isolated reverberation rooms. The sound power incident on the partition under test in the source room is deduced from an estimate of the spatial average of the mean square sound pressure in the room on the assumption that the sound field is diffuse, and the transmitted sound power is determined from a similar measurement in the receiving room where, in addition, the reverberation time must be determined. The sound intensity method has made it possible to measure the transmitted sound power directly using a sound intensity probe. In this case it is not necessary that the sound field in the receiving room is diffuse, which means that only one reverberation room (the source room) is necessary [29]. One cannot measure the incident sound power in the source room using sound intensity, since the method gives the net sound intensity.

The main advantage of the intensity method compared with the conventional approach is that it is possible to evaluate the transmission loss of individual parts of the partition. However, each sound power measurement must obviously satisfy the condition $\Delta_{pI} < \delta_{pIo} - K$.

There are other sources of error than phase mismatch. For example, Roland has called attention to the fact that the traditional method of measuring the sound power transmitted through the partition under test gives the transmitted sound power irrespective of the absorption of the partition, whereas the intensity method gives the *net* power [30]. If a significant part of the absorption in the receiving room is due to the partition then the net power is less than transmitted power. Under such conditions one must increase the absorption of the receiving room; otherwise the intensity method will overestimate the transmission loss because the transmitted sound power is underestimated.

As an interesting by-product of the intensity method it can be mentioned that deviations observed between results determined using the traditional pressure-based method and the intensity method led several authors to re-analyse the traditional method in the 1980s [31, 32] and point out that the Waterhouse correction [33], well established in sound power determination using the reverberation room method (see Section 8.3.1), had been overlooked in the standards for conventional measurements of transmission loss (the ISO 140 series). For more than a decade various authors expressed different opinions about whether the Waterhouse correction should be applied in the source room and not just in the receiving room; see, e.g., [34]. Much later a correction for the extra incident sound power in the source room was derived and it was shown that from a practical point of view it will be cancelled by the Waterhouse correction on the receiving side; in other words that no correction should be applied in conventional pressure-based measurements [35]. However, this cancellation occurs only if the partition is a complete wall.

6.5.4 Measurement of the Emission Sound Pressure Level

The 'emission sound pressure level' is the sound pressure level at an operator's position near a large machine. There is now a standard based on the idea of deducing this level from sound intensity measurements in order to reduce the influence of certain sources of error.

6.6 Sound Absorption

Most materials absorb sound. As we have seen in Chapter 3 we need a precise description of the boundary conditions for solving the wave equation, which leads to a description of material properties in terms of the specific acoustic surface impedance, as described in Section 5.3. However, in many practical applications, for example in architectural acoustics, a simpler measure of the acoustic properties of materials defined in terms of a power ratio, the absorption coefficient, is more useful. By definition the absorption coefficient of a given material is the absorbed fraction of the incident sound power. From this definition it follows that the absorption coefficient takes values between naught and unity. A value of unity implies that all the incident sound power is absorbed.

In general the absorption coefficient of a given material depends on the structure of the sound field (plane wave incidence with a given angle of incidence, for example, or random or diffuse incidence in a room). The absorption for plane waves of normal incidence is studied in Section 7.2.1; the absorption for diffuse sound incidence is studied in Section 8.3.

References

[1] F.J. Fahy: *Sound Intensity* (2nd edition). E&FN Spon, London (1995). (See Chapters 2 and 4.)

[2] P.M. Morse and K.U. Ingard: *Theoretical Acoustics*. Princeton University Press (1984). See Chapter 6.

[3] A.D. Pierce: *Acoustics: An Introduction to Its Physical Principles and Applications*. 2nd edition, Acoustical Society of America, New York (1989). See Section 1–11.

[4] S.W. Rienstra and A. Hirchberg: *An Introduction to Acoustics*. Report 1WDE99-02, Instituut Wiskundige Dienstverlering Eindhoven, Technische Universiteit Eindhoven, The Netherlands (2012). See Section 2.7.

[5] J.A. Mann III, J. Tichy and A.J. Romano: Instantaneous and time-averaged energy transfer in acoustic fields. *Journal of the Acoustical Society of America* **82**, 17–30 (1987).

[6] J.-C. Pascal: Mesure de l'intensité active and réactive dans different champs acoustiques. *Proceedings of International Congress on Recent Developments in Acoustic Intensity Measurement*, pp. 11–19, Senlis, France (1981).

[7] F. Jacobsen: Active and reactive, coherent and incoherent sound fields. *Journal of Sound and Vibration* **130**, 493–507 (1989).

[8] P.J. Westervelt: Acoustical impedance in terms of energy functions. *Journal of the Acoustical Society of America* **23**, 347–349 (1951).

[9] W. Maysenhölder: The reactive intensity of general time-harmonic structure-borne sound fields. *Proceedings of Fourth International Congress on Intensity Techniques*, pp. 63–70, Senlis, France (1993).

[10] F. Jacobsen, A note on instantaneous and time-averaged active and reactive sound intensity. *Journal of Sound and Vibration* **147**, 489–496 (1991).

[11] W.F. Druyvesteyn and H.E. de Bree: A novel sound intensity probe. Comparison with the pair of pressure microphones intensity probes. *Journal of the Audio Engineering Society* **48**, 49–56 (2000).

[12] R. Raangs, W.F. Druyvesteyn and H.-E. de Bree: A low-cost intensity probe. *Journal of the Audio Engineering Society* **51**, 344–357 (2003).

[13] F. Jacobsen and H.-E. De Bree: A comparison of two different sound intensity measurement principles. *Journal of the Acoustical Society of America* **118**, 1510–1517 (2005).

[14] F. Jacobsen and Y. Liu: Near field acoustic holography with particle velocity transducers. *Journal of the Acoustical Society of America* **118**, 3139–3144 (2005).

[15] J. Jacobsen and V. Jaud: Statistically optimized near field acoustic holography using an array of pressure-velocity probes. *Journal of the Acoustical Society of America* **121**, 1550–1558 (2007).

[16] F. Jacobsen and V. Jaud: A note on the calibration of pressure-velocity sound intensity probes. *Journal of the Acoustical Society of America* **120**, 830–837 (2006).

[17] F. Jacobsen, V. Cutanda and P.M. Juhl: A numerical and experimental investigation of the performance of sound intensity probes at high frequencies. *Journal of the Acoustical Society of America* **103**, 953–961 (1998).

[18] J. Pope: Qualifying intensity measurements for sound power determination. *Proceedings of Inter-Noise 89*, pp. 1041–1046, Newport Beach, CA, USA (1989).

[19] U.S. Shirahatti and M.J. Crocker: Two-microphone finite difference approximation errors in the interference fields of point dipole sources. *Journal of the Acoustical Society of America* **92**, 258–267 (1992).

[20] F. Jacobsen, V. Cutanda and P.M. Juhl: A sound intensity probe for measuring from 50 Hz to 10 kHz. *Proceedings of Inter-Noise 96*, pp. 3357–3362, Liverpool, England (1996).

[21] G. Krishnappa: Interference effects in the two-microphone technique of acoustic intensity measurements. *Noise Control Engineering Journal* **21**, 126–135 (1983).

[22] F. Jacobsen: A simple and effective correction for phase mismatch in intensity probes. *Applied Acoustics* **33**, 165−180 (1991).

[23] M. Ren and F. Jacobsen: Phase mismatch errors and related indicators in sound intensity measurement. *Journal of Sound and Vibration* **149**, 341−347 (1991).

[24] F. Jacobsen, Sound field indicators: Useful tools. *Noise Control Engineering Journal* **35**, 37−46 (1990).

[25] F. Jacobsen and E.S. Olsen: Testing sound intensity probes in interference fields. *Acustica* **80**, 115−126 (1994).

[26] F. Jacobsen: Sound intensity measurements at low levels. *Journal of Sound and Vibration* **166**, 195−207 (1993).

[27] M.J. Crocker: Sound power determination from sound intensity – To scan or not to scan. *Noise Control Engineering Journal* **27**, 67 (1986).

[28] U.S. Shirahatti and M.J. Crocker: Studies of the sound power estimation of a noise source using the two-microphone sound intensity technique. *Acustica* **80**, 378−387 (1994).

[29] M.J. Crocker, P.K. Raju and B. Forssen, Measurement of transmission loss of panels by the direct determination of transmitted acoustic intensity. *Noise Control Engineering Journal* **17**, 6−11 (1981).

[30] J. Roland and C. Martin and M. Villot: Room to room transmission: What is really measured by intensity? *Proceedings of 2nd International Congress on Acoustic Intensity*, pp. 539−546, Senlis, France (1985).

[31] R.E. Halliwell and A.C.C. Warnock: Sound transmission loss: Comparison of conventional techniques with sound intensity techniques. *Journal of the Acoustical Society of America* **77**, 2094−2103 (1985).

[32] B.G. van Zyl, P.J. Erasmus and F. Anderson: On the formulation of the sound intensity method for determining sound reduction indices. *Applied Acoustics* **22**, 213−228 (1987).

[33] R.V. Waterhouse: Interference patterns in reverberant sound fields. *Journal of the Acoustical Society of America* **27**, 247−258 (1955).

[34] M. Vorländer: Revised relation between the sound power and the average sound pressure level in rooms and consequences for acoustic measurements. *Acustica* **81**, 332−343 (1995).

[35] F. Jacobsen and E. Tiana-Roig: Measurement of the sound power incident on the walls of a reverberation room with near field acoustic holography. *Acta Acustica united with Acustica* **96**, 76−91 (2010).

7

Duct Acoustics

7.1 Introduction

Duct acoustics is the branch of acoustics concerned with sound propagation in pipes and ducts. Sound propagation within ducts is of practical concern in many areas of acoustics and noise control. Cylindrical cavities are used for testing or calibrating transducers, and impedance tubes are used for measuring the acoustic properties of absorbing materials. In a number of musical instruments (brass and woodwind instruments, organs) the sound is generated in pipes. Air-distributing systems for heating, ventilating and air-conditioning also transmit noise. Exhaust noise from petrol- and diesel-driven reciprocating engines must be reduced by silencers, which essentially are combinations of pipes.

The purpose of this chapter is to introduce the topic. Non-linear effects are not dealt with, and the effects of viscosity, heat conduction and mean flow are only touched upon.

7.2 Plane Waves in Ducts with Rigid Walls

The sound field in a uniform duct with rigid walls, sometimes referred to as an acoustic transmission line, is fairly simple in form at low frequencies where the cross-sectional dimensions are much smaller than the wavelength. As will be shown later, below a certain frequency that depends on the cross-sectional shape and dimensions of the duct only plane waves can propagate, from which it follows that the sound field is essentially one-dimensional. This is the case in many situations of practical interest.

7.2.1 The Sound Field in a Tube Terminated by an Arbitrary Impedance

In the following it is assumed that the sound field is one-dimensional, just as in Section 3.1. Consider a duct terminated at $z = 0$ by a surface with a certain acoustic impedance, $Z_{a,t}$, as sketched in Figure 7.1. If a sound field is generated by a source, say, a piston, at a position $z < 0$, then the first term of Equation (3.11) can be interpreted as the incident part of the sound field, and the second term is the reflected part. It is obvious that $|p_-| \leq |p_+|$. The complex ratio of the amplitudes at $z = 0$ is the reflection factor, R,

Fundamentals of General Linear Acoustics, First Edition. Finn Jacobsen and Peter Møller Juhl.
© 2013 John Wiley & Sons, Ltd. Published 2013 by John Wiley & Sons, Ltd.

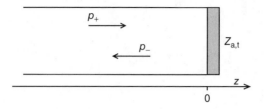

Figure 7.1 A tube terminated by a surface with the acoustic impedance $Z_{a,t}$

which completely characterises the acoustic properties of the surface for normal sound incidence; cf. Equation (3.15).

Expressed in terms of the reflection factor the sound pressure and the axial component of the particle velocity become

$$\hat{p} = p_+(e^{-jkz} + Re^{jkz})e^{j\omega t}. \tag{7.1}$$

$$\hat{u}_z = \frac{p_+}{\rho c}(e^{-jkz} - Re^{jkz})e^{j\omega t}. \tag{7.2}$$

The wave impedance is

$$Z_s(z) = \frac{\hat{p}}{\hat{u}_z} = \rho c \frac{e^{-jkz} + Re^{jkz}}{e^{-jkz} - Re^{jkz}}. \tag{7.3}$$

In Figure 7.2 are shown the modulus and phase of the wave impedance as functions of the position for two different values of the standing wave ratio s (cf. Equation (3.18)).

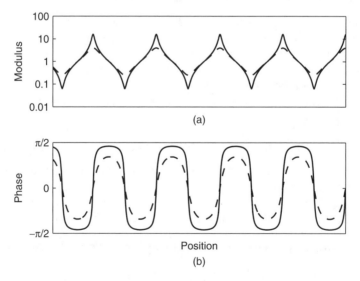

(a)

(b)

Figure 7.2 (a) Modulus (in ρc-units) and (b) phase of the wave impedance in a one-dimensional sound field for two different values of the standing wave ratio: solid line, 24 dB; dashed line, 12 dB

The boundary condition at $z = 0$ implies that

$$\frac{\hat{p}(0)}{S\hat{u}_z(0)} = \frac{Z_s(0)}{S} = Z_{a,t}. \tag{7.4}$$

It now follows that

$$Z_{a,t} = \frac{\rho c}{S} \frac{1 + R}{1 - R}, \tag{7.5}$$

which leads to

$$R = \frac{Z_{a,t} - \rho c / S}{Z_{a,t} + \rho c / S}. \tag{7.6}$$

Equation (7.6) demonstrates that the quantity $\rho c / S$ is important; this quantity may be called the characteristic impedance of the tube. Apparently, there is total reflection in phase ($R \simeq 1$) when the terminating surface is almost rigid ($|Z_{a,t}| \gg \rho c / S$), there is total reflection in antiphase ($R \simeq -1$) when the terminating impedance is very small ($|Z_{a,t}| \ll \rho c / S$), and there is no reflection ($R = 0$) when the terminating impedance equals the characteristic impedance of the tube, $\rho c / S$.

The sound intensity is

$$I_z(z) = \frac{1}{2} \mathrm{Re}\left\{ \hat{p}(z)\, \hat{u}_z^*(z) \right\} = \frac{1}{2} \mathrm{Re}\left\{ p_+ \left(e^{-jkz} + Re^{jkz} \right) \frac{p_+^*}{\rho c} \left(e^{jkz} - R^* e^{-jkz} \right) \right\}$$
$$= \frac{|p_+|^2}{2\rho c}(1 - |R|^2) = \frac{p_{\max} p_{\min}}{2\rho c}, \tag{7.7}$$

cf. Equations (3.16) and (3.17). Equation (7.7) shows that the axial sound intensity is independent of the position, as it must be in a one-dimensional sound field in a tube with rigid walls because of conservation of power.

Example 7.1 Normal incidence absorption: By definition, the absorption coefficient of the material that terminates the tube is the ratio of absorbed to incident sound power; cf. Section 5.3. Evidently, this quantity must take values between zero and unity. The incident sound power is the sound power associated with the incident wave,

$$P_+ = SI_+ = S\frac{|p_+|^2}{2\rho c},$$

and since the sound power absorbed by the material is

$$P_a = SI_z = S\frac{|p_+|^2}{2\rho c}(1 - |R|^2),$$

it follows that the absorption coefficient is

$$\alpha = I_z / I_+ = 1 - |R|^2.$$

The modulus of the reflection factor can be expressed in terms of the standing wave ratio,

$$|R| = \frac{s - 1}{s + 1},$$

cf. Equation (3.19), from which it can be seen that

$$\alpha = 1 - \left| \frac{s-1}{s+1} \right|^2 = \frac{4s}{(1+s)^2}.$$

The standing wave ratio, and thus the normal incidence absorption coefficient of the material at the end of the tube, can be measured by drawing a pressure microphone through the tube and recording the sound pressure as a function of the axial position. This is a classical method of determining absorption coefficients. Figure 7.3 shows examples of standing wave patterns with different terminations of the tube.

Example 7.2 The transfer function method of measuring absorption: It is possible to determine the reflection factor R and thus the absorption coefficient of the material at the end of the tube by measuring the transfer function between the signals from two pressure microphones mounted in the wall of the tube. Let

$$\hat{p}_1 = \hat{p}(z + \Delta z/2), \quad \hat{p}_2 = \hat{p}(z - \Delta z/2),$$

where Δz is the separation distance (see Figure 7.4). The transfer function between these signals is

$$H_{12} = \frac{\hat{p}_2}{\hat{p}_1} = \frac{e^{-jk(z-\Delta z/2)} + Re^{jk(z-\Delta z/2)}}{e^{-jk(z+\Delta z/2)} + Re^{jk(z+\Delta z/2)}}.$$

Rearranging gives the expression

$$R = e^{-j2kz} \frac{e^{jk\Delta z} - H_{12}}{H_{12}e^{jk\Delta z} - 1},$$

which shows that R can be calculated from H_{12}, k, z and Δz unless the distance between the microphones is a multiple of half a wavelength. Since this method does not require a traversing microphone, it is generally much faster than the classical method based on measuring the standing wave ratio. In practice the tube is driven with wide-band random noise, and the transfer function between the two pressure signals is measured using a multi-channel FFT analyser. By contrast the classical method requires driving the tube with one frequency at a time.

Equations (7.1) and (7.2) make it possible to determine the acoustic input impedance of the tube,

$$Z_{a,i} = \frac{\hat{p}(-l)}{S\hat{u}_z(-l)} = \frac{\rho c}{S} \frac{e^{jkl} + Re^{-jkl}}{e^{jkl} - Re^{-jkl}}, \tag{7.8}$$

where l is the length of the tube. Combining this with Equation (7.6) gives the acoustic input impedance expressed in terms of the acoustic impedance of the material at the end

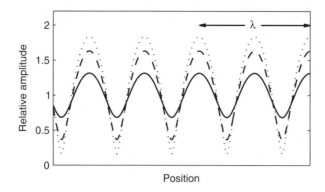

Figure 7.3 Amplitude in standing wave pattern relative to amplitude of incident wave in a tube terminated by a sample of a material with an absorption coefficient of 90% (solid line), 60% (dashed line) and 30% (dotted line)

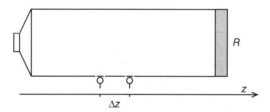

Figure 7.4 The 'two-microphone' technique of measuring absorption coefficients in a tube

of the tube:

$$
Z_{\mathrm{a,i}} = \frac{\rho c}{S} \frac{(Z_{\mathrm{a,t}} + \rho c/S)\mathrm{e}^{\mathrm{j}kl} + (Z_{\mathrm{a,t}} - \rho c/S)\mathrm{e}^{-\mathrm{j}kl}}{(Z_{\mathrm{a,t}} + \rho c/S)\mathrm{e}^{\mathrm{j}kl} - (Z_{\mathrm{a,t}} - \rho c/S)\mathrm{e}^{-\mathrm{j}kl}}
$$
$$
= \frac{Z_{\mathrm{a,t}} \cos kl + \mathrm{j}(\rho c/S) \sin kl}{\mathrm{j}(\rho c/S)Z_{\mathrm{a,t}} \sin kl + \cos kl}. \tag{7.9}
$$

It is apparent that the input impedance varies with the frequency unless $Z_{\mathrm{a,t}} = \rho c/S$. Note also that $Z_{\mathrm{a,i}} = Z_{\mathrm{a,t}}$ when kl is a multiple of π corresponding to the length of the tube being a multiple of half a wavelength. It is also interesting to observe that

$$
Z_{\mathrm{a,i}} = \left(\frac{\rho c}{S}\right)^2 \frac{1}{Z_{\mathrm{a,t}}} \tag{7.10}
$$

when kl is an odd-numbered multiple of $\pi/2$ corresponding to the length of the tube being an odd-numbered multiple of a quarter of a wavelength; a tube of this length transforms a large impedance into a small one and *vice versa*.

Figure 7.5 shows an example of a measured acoustic input impedance of a tube terminated by a rigid cap. The sound pressure and the particle velocity have been measured using a sound intensity probe. The measurement is not reliable below 100 Hz.

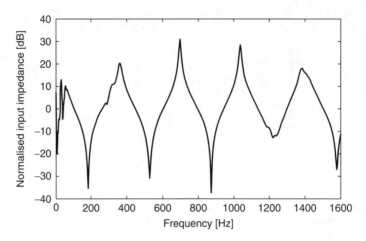

Figure 7.5 Modulus (in ρc-units) of the specific input impedance of a 50-cm long tube terminated by a rigid cap, measured using a sound intensity probe

When the terminating impedance is large ($|Z_{a,t}| \gg \rho c/S$), the acoustic input impedance of the tube becomes very small when l equals an odd-numbered multiple of a quarter of a wavelength (cf. Example 3.2). The corresponding discrete frequencies are the *resonance frequencies* of the system; the resonance frequencies are odd-numbered harmonics of the fundamental. Conversely, the input impedance becomes very large when l equals a multiple of half a wavelength, which occurs at the *antiresonance* frequencies of the system.

One way of generating sound in the tube is by placing a vibrating piston at one end. Let the source be a surface at $z = -l$ vibrating with the velocity $U e^{j\omega t}$. In the frequency range we are concerned with, the decisive property of the source is its volume velocity

$$Q = \int_S U(x, y)\mathrm{d}S. \tag{7.11}$$

The boundary condition at $z = -l$ implies that the volume velocity in the sound field is

$$\hat{q}(-l) = S\hat{u}(-l) = Q e^{j\omega t}. \tag{7.12}$$

Example 7.3 A closed tube driven by a piston: A tube of the length l terminated by a rigid surface at $z = 0$ is driven at $z = -l$ by a piston with the vibrational velocity $U e^{j\omega t}$. Evidently

$$\hat{u}_z(-l) = U e^{j\omega t},$$

and since, from Equation (3.14),

$$\hat{u}_z(l) = \frac{2\mathrm{j}p_+}{\rho c} \sin kl \ e^{j\omega t},$$

it follows that

$$p_+ = \frac{U\rho c}{2\mathrm{j} \sin kl},$$

from which it can be seen that

$$\hat{p}(z) = -\frac{jU\rho c \cos kz}{\sin kl} e^{j\omega t}.$$

Note that the sound pressure goes towards infinity at the antiresonance frequencies owing to the fact that losses have been ignored. In practice there will always be some losses, and the source will have a finite acoustic impedance, which means that it cannot maintain its volume velocity if the sound pressure on its surface goes towards infinity.

If the tube is driven with a small piston that is mounted in the wall, say at $z = z_0$, the boundary conditions at $z = z_0$ are

$$\hat{p}(z_0^+) = \hat{p}(z_0^-) \tag{7.13}$$

and

$$\hat{q}(z_0^+) - \hat{q}(z_0^-) = Q e^{j\omega t}, \tag{7.14}$$

where z_0^+ and z_0^- are the limits $z_0 \pm \alpha$ with $\alpha \to 0$, and $Q e^{j\omega t}$ is the volume velocity of the piston. The discontinuity in the volume velocity is due to the fact that the particle velocity is a vector.

Example 7.4 An infinite tube driven by a piston mounted in the wall: A piston with the volume velocity $Q e^{j\omega t}$ is mounted at $z = 0$ in the wall of infinite tube. This source will generate plane waves travelling to the left and to the right:

$$\left. \begin{array}{l} \hat{p}(z) = p_+ e^{j(\omega t - kz)} \\ \hat{u}_z(z) = \dfrac{p_+}{\rho c} e^{j(\omega t - kz)} \end{array} \right\} \text{ for } z > 0,$$

and

$$\left. \begin{array}{l} \hat{p}(z) = p_- e^{j(\omega t + kz)} \\ \hat{u}_z(z) = -\dfrac{p_-}{\rho c} e^{j(\omega t + kz)} \end{array} \right\} \text{ for } z < 0,$$

From Equations (7.13) and (7.14) it follows that $p_+ = p_- = (\rho c / S)(Q/2)$. In effect the right- and left-going waves 'share' the volume velocity injected at $z = 0$.

Example 7.5 A semi-infinite tube driven by a piston mounted in the wall: A semi-infinite tube is driven by a small loudspeaker with a volume velocity of $Q e^{j\omega t}$ mounted in the wall the distance l from a rigid termination (see Figure 7.6). At positions between the source and the rigid surface ($0 < z < l$),

$$\hat{p}(z) = A\mathrm{e}^{\mathrm{j}(\omega t - kz)} + B\mathrm{e}^{\mathrm{j}(\omega t + kz)} \quad \text{and} \quad \hat{q}_z(z) = \frac{S}{\rho c}\left(A\mathrm{e}^{\mathrm{j}(\omega t - kz)} - B\mathrm{e}^{\mathrm{j}(\omega t + kz)}\right),$$

and since $q(0) = 0$ it can be seen that $A = B$. To the right of the source (at $z > 1$),

$$\hat{p}(z) = C\mathrm{e}^{\mathrm{j}(\omega t - kz)} \quad \text{and} \quad \hat{q}_z(z) = \frac{S}{\rho c}C\mathrm{e}^{\mathrm{j}(\omega t - kz)}.$$

From Equation (7.13),

$$2A \cos kl = C\mathrm{e}^{-\mathrm{j}kl},$$

and from Equation (7.14),

$$\frac{S}{\rho c}C\mathrm{e}^{-\mathrm{j}kl} + 2\mathrm{j}\frac{S}{\rho c}A \sin kl = Q,$$

from which it follows that the sound pressure between the source and the rigid surface is

$$\hat{p}(z) = Q\frac{\rho c}{S} \cos kz \, \mathrm{e}^{-\mathrm{j}kl}\mathrm{e}^{\mathrm{j}\omega t} \quad \text{for } 0 \le z \le l$$

(a standing wave), and

$$\hat{p}(z) = Q\frac{\rho c}{S} \cos kl \, \mathrm{e}^{\mathrm{j}(\omega t - kz)} \quad \text{for } z \ge l$$

(a propagating wave) to the right of the source. Note that $p(z) = 0$ for $z > l$ at frequencies where l equals an odd-numbered multiple of a quarter of a wavelength. The physical explanation is that the sound field to the right of the source is the sum of the direct wave from the source and the left-going wave, reflected at $z = 0$,

$$\hat{p}(z) = Q\frac{\rho c}{2S}\mathrm{e}^{\mathrm{j}(\omega t - k(z-l))} + Q\frac{\rho c}{2S}\mathrm{e}^{\mathrm{j}(\omega t - k(z+l))}.$$

When l is an odd-numbered multiple of a quarter of a wavelength, the reflected wave has travelled an odd-numbered multiple of half a wavelength longer than the direct wave and is in antiphase; therefore it cancels the direct wave.

7.2.2   Radiation of Sound from an Open-ended Tube

An open-ended tube is in effect terminated by the radiation impedance, which is the impedance that would load a piston placed at the end of the tube. The radiation impedance of a flanged tube (a tube with the open end in a large, rigid plane) is identical with the radiation impedance of a piston set in a large, plane, rigid baffle; the radiation impedance of an unflanged tube is similar to the radiation impedance of a small monopole in free space. As we shall see, at very low frequencies these quantities are much smaller than the characteristic impedance of the tube, which, with Equation (7.6), leads to the conclusion that $R \simeq -1$; there is a strong reflection in antiphase. Accordingly, the input impedance

Figure 7.6   A semi-infinite tube driven by a loudspeaker at an arbitrary position

at the other end of the tube exhibits minima when the (effective) length of the tube is a multiple of half a wavelength.

It can be shown that the radiation impedance of a circular piston of the radius a mounted in an infinite baffle is

$$Z_{a,r} = \frac{\rho c}{S} \left(1 - \frac{J_1(2ka)}{ka} + j\frac{H_1(2ka)}{ka} \right), \tag{7.15}$$

where J_1 is the Bessel function and H_1 is the Struve function of first order [1]. At low frequencies this expression can be approximated as follows,

$$Z_{a,r} \simeq \frac{\rho c}{S} \left(\frac{1}{2}(ka)^2 + j\frac{8}{3\pi}ka \right). \tag{7.16}$$

The approximation is fair if $ka < 0.5$.

At very low frequencies a considerable part of the incident sound power is reflected from the open end of the tube, as mentioned above. The *transmission coefficient* describes the small fraction of the incident sound power radiated into the open. From Equation (7.6) we have

$$R = \frac{Z_{a,r} - \rho c/S}{Z_{a,r} + \rho c/S} \simeq - \frac{1 - \frac{1}{2}(ka)^2 - j\frac{8}{3\pi}ka}{1 + \frac{1}{2}(ka)^2 + j\frac{8}{3\pi}ka}, \tag{7.17}$$

from which it follows that the transmission coefficient τ is given by

$$\tau = 1 - |R|^2 \simeq 1 - \frac{\left(1 - \frac{(ka)^2}{2}\right)^2 + \left(\frac{8}{3\pi}\right)^2 (ka)^2}{\left(1 + \frac{(ka)^2}{2}\right)^2 + \left(\frac{8}{3\pi}\right)^2 (ka)^2} \simeq 2(ka)^2. \tag{7.18}$$

Note that the transmission coefficient to this approximation is independent of the imaginary part of the radiation impedance, even though the imaginary part is much larger than the real part. The imaginary part of the radiation impedance corresponds to the impedance of a mass of a volume of the medium of the size $S\Delta 1$, where

$$\Delta l = \frac{8a}{3\pi} \tag{7.19}$$

is the *end correction*. This becomes apparent if the radiation impedance is written as a *mechanical* impedance (the ratio of force to velocity):

$$Z_{m,r} = \frac{\hat{F}}{\hat{u}} = \frac{S\hat{p}}{\hat{q}/S} = S^2 Z_{a,r} \simeq \rho c S \left(\frac{1}{2}(ka)^2 + jk\,\Delta l\right) = \frac{1}{2}\rho c S (ka)^2 + j\omega\rho S\,\Delta l. \quad (7.20)$$

The main effect of the imaginary part of the radiation impedance is to shift the position where reflection in antiphase occurs to a virtual plane outside the tube. This can be seen as follows. If we ignore the small real part of the radiation impedance Equation (7.17) becomes

$$R \simeq -\frac{1 - j\dfrac{8}{3\pi}ka}{1 + j\dfrac{8}{3\pi}ka} \simeq -\frac{e^{-jk\,\Delta l}}{e^{jk\,\Delta l}} = -e^{-j2k\,\Delta l}, \quad (7.21)$$

corresponding to the input impedance of the tube being

$$Z_{a,i} \simeq \frac{\rho c}{S}\frac{e^{jkl} - e^{-jkl}e^{-j2k\,\Delta l}}{e^{jkl} + e^{-jkl}e^{-j2k\,\Delta l}} = \frac{\rho c}{S}\frac{e^{jk(l+\Delta l)} - e^{-jk(l+\Delta l)}}{e^{jk(l+\Delta l)} + e^{-jk(l+\Delta l)}} = j\frac{\rho c}{S}\tan k(l + \Delta l), \quad (7.22)$$

and this is exactly the result we would obtain if a tube of the length $l + \Delta l$ were terminated by a surface with an acoustic impedance of zero (cf. Equation (7.9) with $Z_{a,t} = 0$). Equation (7.22) explains why the quantity Δl is called the end correction: the 'effective length' of the tube is $l + \Delta l$. Ignoring the real part of the radiation impedance is too coarse an approximation at the resonance frequencies, though. At these frequencies

$$Z_{a,i} \simeq \frac{\rho c}{S}\frac{(ka)^2}{2}, \quad (7.23)$$

which is small but not zero as implied by Equation (7.22). In a similar manner it can be shown that

$$Z_{a,i} \simeq \frac{\rho c}{S}\frac{2}{(ka)^2}, \quad (7.24)$$

at the antiresonance frequencies, and this impedance is large but not infinite as implied by Equation (7.22). In fact, the tube behaves as if it had a length of $l + \Delta l$ and were terminated by a small, purely real impedance, $\mathrm{Re}\{Z_{a,t}\}$.

There is no simple way of calculating the radiation impedance of a thin-walled tube that radiates into open space without a flange. However, it can be shown [2] that in the low-frequency limit,

$$Z_{a,r} \simeq \frac{\rho c}{S}\left(\frac{(ka)^2}{4} + j0.61\cdot ka\right), \quad (7.25)$$

corresponding to a transmission coefficient of

$$\tau \simeq (ka)^2 \quad (7.26)$$

and an end correction that takes a value of

$$\Delta l \simeq 0.61\cdot a. \quad (7.27)$$

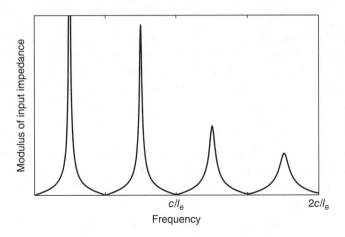

Figure 7.7 The input impedance of an open-ended tube

Apparently, a large flange tends to double the real part of the radiation impedance – in agreement with the fact that the solid angle is (almost) halved by the flange, which implies a better impedance match. It is also worth noting that the real part of the radiation impedance of the unflanged tube,

$$\text{Re}\{Z_{\text{a,r}}\} \simeq \frac{\rho c}{\pi a^2}\frac{(ka)^2}{4} = \frac{\rho c k^2}{4\pi} = \frac{\rho \omega^2}{4\pi c}, \tag{7.28}$$

is identical with the radiation resistance of a small pulsating sphere, as we shall see in Section 9.4.

Figure 7.7 shows the input impedance of an open-ended tube. As the frequency is increased the impedance peaks and troughs become less and less pronounced because the real part of the radiation impedance is increased.

Example 7.6 A source of high acoustic impedance driving an open tube: A source of high acoustic impedance, that is, a source that tends to maintain its volume velocity irrespective of the acoustic load, tends to excite the antiresonance frequencies of the tube. The generator in reed instruments, e.g. clarinets and oboes, and brass instruments, e.g. trumpets, may be regarded as a source of high impedance; therefore such instruments play at impedance peaks (antiresonance frequencies), which are odd-numbered harmonics of the fundamental. A tone produced by a clarinet therefore has a strong content of odd-numbered harmonics [3].

Example 7.7 A source of low acoustic impedance driving an open tube: A source of low acoustic impedance, that is, a source that tends to produce a given sound pressure at its surface irrespective of the acoustic load, tends to excite the resonance frequencies of the tube. The generator in flutes and organ pipes may be regarded as a source of low impedance; therefore flutes and open organ pipes play at impedance minima (resonance frequencies), which occur at all harmonics. It follows that a tone produced by a flute has even as well as odd harmonics [3].

Example 7.8 A closed organ pipe: A tone produced by a *closed* organ pipe has only odd-numbered harmonics, because the resonances of a closed tube occur when the length equals an odd-numbered multiple of a quarter of a wavelength; cf. Examples 7.6 and 7.7.

7.3   Sound Transmission Through Coupled Pipes

A silencer (or muffler) is a device designed to reduce the emission of noise from a pipe, for example the noise from the exhaust of an engine. Reactive silencers are composed of coupled tubes without absorbing material. They work by reflecting the sound from the source. Dissipative silencers absorb acoustic energy.

There is no simple way of dealing with the sound field in a system of coupled tubes as the one sketched in Figure 7.8 in the general case. However, at low frequencies where the sound field in each component part can be assumed to be one-dimensional, the problem is a fairly simple one. At each junction it can be assumed that the output sound pressure and volume velocity of one system equal the input sound pressure and volume velocity of the adjoining system; therefore the acoustic input impedance of the exit system is the acoustic load on the preceding system.

Example 7.9 Reflection from a sudden change in cross sectional area: Two tubes with cross-sectional areas S_1 and S_2 are connected, as sketched in Figure 7.9. If the second tube is infinite, its input impedance is $\rho c / S_2$, from which it follows that the reflection factor at the junction seen from the first tube is

$$R = \frac{\rho c / S_2 - \rho c / S_1}{\rho c / S_2 + \rho c / S_1} = \frac{S_1 - S_2}{S_1 + S_2}$$

(cf. Equation (7.6)). Note that $R \simeq 1$ if $S_2 \ll S_1$, in accordance with the fact that a sudden, strong area contraction almost acts as a rigid wall. Conversely, $R \simeq -1$ if $S_2 \gg S_1$; in this case the junction acts almost as a pressure-release surface.

The transmission coefficient is

$$\tau = 1 - |R|^2 = \frac{4 S_1 S_2}{(S_1 + S_2)^2}.$$

It can be seen that the fraction of incident sound power transmitted from one tube to another is the same in both directions.

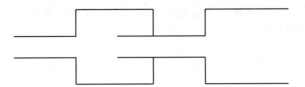

Figure 7.8 A system of coupled pipes

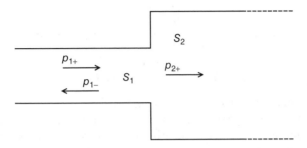

Figure 7.9 Transmission and reflection at an area discontinuity between two coupled pipes

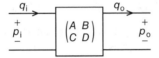

Figure 7.10 An acoustic two-port

It is physically obvious that it is unrealistic to expect the sound field to be strictly one-dimensional in the immediate vicinity of a discontinuity as the one shown in Figure 7.9. Nevertheless, if one's main concern is the transmission properties of the composite system it is usually a good approximation to assume that the sound field is one-dimensional (below a certain frequency determined by the cross-sectional dimensions), because the local disturbances correspond to higher-order modes that die out exponentially with the distance and have little effect on the overall transmission. Higher-order modes are analysed in Section 7.5.

7.3.1 The Transmission Matrix

The low-frequency approximation implies that each subsystem is an acoustic two-port (or four-pole system) with two (and only two) unknown parameters, the complex amplitudes of two interfering waves travelling in opposite directions. Such a system can be described by its transmission (or four-pole) matrix, as follows,

$$\begin{pmatrix} \hat{p}_i \\ \hat{q}_i \end{pmatrix} = \begin{pmatrix} A & B \\ C & D \end{pmatrix} \begin{pmatrix} \hat{p}_o \\ \hat{q}_o \end{pmatrix} \tag{7.29}$$

where $\hat{p}_i$, $\hat{p}_o$, $\hat{q}_i$ and $\hat{q}_o$ are the sound pressures and volume velocities at the input and at the output; see also Figure 7.10. It is apparent that

$$A = \frac{\hat{p}_i}{\hat{p}_o}\bigg|_{\hat{q}_o=0} , \quad B = \frac{\hat{p}_i}{\hat{q}_o}\bigg|_{\hat{p}_o=0} , \quad C = \frac{\hat{q}_i}{\hat{p}_o}\bigg|_{\hat{q}_o=0} , \quad D = \frac{\hat{q}_i}{\hat{q}_o}\bigg|_{\hat{p}_o=0} , \quad (7.30a, 7.30b, 7.30c, 7.30d)$$

The four-pole parameters are not independent. From the reciprocity principle it can be shown that

$$C = \frac{\hat{q}_i}{\hat{p}_o}\bigg|_{\hat{q}_o=0} = -\frac{\hat{q}_o}{\hat{p}_i}\bigg|_{\hat{q}_i=0}, \quad (7.31)$$

where the latter quantity can be deduced from the matrix equation

$$\begin{pmatrix} \hat{p}_i \\ 0 \end{pmatrix} = \begin{pmatrix} A & B \\ C & D \end{pmatrix} \begin{pmatrix} \hat{p}_o \\ \hat{q}_o \end{pmatrix}. \quad (7.32)$$

This leads to the conclusion that the matrix determinant $AD - BC$ equals unity.

The description of the subsystems in terms of their four-pole parameters is very convenient, because the output of one system is the input of the next system, from which it follows that the transmission matrix of a system consisting of cascaded subsystems is obtained by matrix multiplication,

$$\begin{pmatrix} \hat{p}_{i,1} \\ \hat{q}_{i,1} \end{pmatrix} = \begin{pmatrix} A_1 & B_1 \\ C_1 & D_1 \end{pmatrix} \begin{pmatrix} A_2 & B_2 \\ C_2 & D_2 \end{pmatrix} \cdots \begin{pmatrix} A_m & B_m \\ C_m & D_m \end{pmatrix} \begin{pmatrix} \hat{p}_{o,m} \\ \hat{q}_{o,m} \end{pmatrix} = \begin{pmatrix} A & B \\ C & D \end{pmatrix} \begin{pmatrix} \hat{p}_{o,m} \\ \hat{q}_{o,m} \end{pmatrix}. \quad (7.33)$$

Figure 7.11 illustrates a composite system and its transmission matrix based on the product of transmission matrices of subsystems.

The source that generates the noise that the silencer is supposed to reduce can be described either in terms of its volume velocity 'in parallel with' its acoustic impedance or its sound pressure 'in series with' its acoustic impedance; in other words in terms of equivalent electrical circuits; see Figure 7.12. The former description implies that the input volume velocity at the inlet of the system differs from the unloaded volume velocity

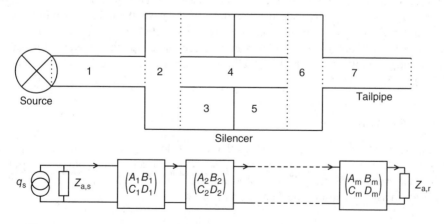

Figure 7.11 A system composed of coupled pipes and its transmission matrix

Figure 7.12   Two equivalent electrical network descriptions of a source

of the source, $\hat{q}_s$, because of the acoustic load of the input system,

$$\hat{q}_i = \hat{q}_s \frac{Z_{a,s}}{Z_{a,s} + Z_{a,i}},\tag{7.34}$$

unless $|Z_{a,s}| \gg |Z_{a,i}|$; it also shows that the actual input sound pressure is

$$\hat{p}_i = \hat{q}_s \frac{Z_{a,s} Z_{a,i}}{Z_{a,s} + Z_{a,i}}.\tag{7.35}$$

The latter description entails that

$$\hat{p}_i = \hat{p}_s \frac{Z_{a,i}}{Z_{a,s} + Z_{a,i}},\tag{7.36}$$

which shows that the actual input sound pressure differs from $\hat{p}_s$ unless $|Z_{a,s}| \ll |Z_{a,i}|$. It also implies that

$$\hat{q}_i = \frac{\hat{p}_s}{Z_{a,s} + Z_{a,i}}.\tag{7.37}$$

The two source descriptions are equivalent. However, the former description is the natural choice if the volume velocity of the source is almost independent of the acoustic load, and the latter description is more appropriate if the sound generation mechanism tends to keep the sound pressure constant. Often the input impedance of the system driven by the source varies strongly with the frequency (cf. Figures 7.5 and 7.7), which means that it is necessary to take account of the finite source impedance, though.[1]

It is a simple matter to derive the transmission matrix of a cylindrical tube with the length l and the cross-sectional area S. With an appropriate choice of the local coordinate system we can write

$$\hat{p}_i = p_+ e^{j(\omega t + kl)} + p_- e^{j(\omega t - kl)},\tag{7.38}$$

$$\hat{q}_i = \frac{S}{\rho c} \left(p_+ e^{j(\omega t + kl)} - p_- e^{j(\omega t - kl)} \right),\tag{7.39}$$

and

$$\hat{p}_o = (p_+ + p_-) e^{j\omega t},\tag{7.40}$$

$$\hat{q}_o = \frac{S}{\rho c} (p_+ - p_-) e^{j\omega t},\tag{7.41}$$

---

[1] Outside of ducts and small cavities most sources of air-borne noise are able to maintain their volume velocity irrespective of the radiation load. This corresponds to the case in which $|Z_{a,s}| \gg |Z_{a,r}|$.

for the input and output quantities. The four-pole parameters now follow from Equations (7.30a–d). For example, a rigid termination ($\hat{q}_o = 0$) leads to the requirement that $p_+ = p_-$, from which we conclude that

$$A = \left.\frac{\hat{p}_i}{\hat{p}_o}\right|_{\hat{q}_o=0} = \cos kl. \tag{7.42}$$

The resulting transmission matrix of the tube is

$$\begin{pmatrix} A & B \\ C & D \end{pmatrix} = \begin{pmatrix} \cos kl & j\dfrac{\rho c}{S}\sin kl \\ j\dfrac{S}{\rho c}\sin kl & \cos kl \end{pmatrix}. \tag{7.43}$$

Example 7.10 The transmission matrix of two tubes with the same cross section: A tube of length l might be regarded as two coupled tubes of lengths l_1 and l_2 where $l_1 + l_2 = 1$, in agreement with the fact that

$$\begin{pmatrix} \cos kl_1 & j\dfrac{\rho c}{S}\sin kl_1 \\ j\dfrac{S}{\rho c}\sin kl_1 & \cos kl_1 \end{pmatrix}\begin{pmatrix} \cos kl_2 & j\dfrac{\rho c}{S}\sin kl_2 \\ j\dfrac{S}{\rho c}\sin kl_2 & \cos kl_2 \end{pmatrix}$$

$$= \begin{pmatrix} \cos k\left(l_1 + l_2\right) & j\dfrac{\rho c}{S}\sin k(l_1 + l_2) \\ j\dfrac{S}{\rho c}\sin k(l_1 + l_2) & \cos k(l_1 + l_2) \end{pmatrix}.$$

The input impedance of a system terminated by a given impedance $Z_{a,t}$ follows from Equation (7.29):

$$Z_{a,i} = \frac{AZ_{a,t} + B}{CZ_{a,t} + D}. \tag{7.44}$$

The sound pressure and volume velocity at the termination impedance of a system driven by a given source follow from Equations (7.29), (7.34) and (7.44),

$$\hat{p}_o = \hat{q}_s \frac{Z_{a,s}Z_{a,t}}{AZ_{a,t} + B + Z_{a,s}(CZ_{a,t} + D)}, \tag{7.45a}$$

$$\hat{q}_o = \hat{q}_s \frac{Z_{a,s}}{AZ_{a,t} + B + Z_{a,s}(CZ_{a,t} + D)}, \tag{7.45b}$$

which demonstrate the influence of the source impedance.

Example 7.11 Calculation of input impedance using the transmission matrix: Some of the results derived in Section 7.2 can be derived more easily if use is made of the transmission matrix of a tube. For example, the input impedance of a tube terminated by a given impedance, Equation (7.9), now follows immediately from Equation (7.44).

The formalism derived in the foregoing makes it a simple matter to calculate the sound pressure and volume velocity at the junction between two subsystems. It is apparent from Figure 7.13 that

$$\hat{p}_x = \hat{p}_o(A' + B'/Z_{a,t}), \tag{7.46a}$$

$$\hat{q}_x = \hat{p}_o(C' + D'/Z_{a,t}), \tag{7.46b}$$

where p_x and q_x are the sound pressure and volume velocity at the input of the terminating subsystem with four-pole parameters A', B', C' and D' (which itself may be composed of several subsystems). Combining Equations (7.45) and (7.46) gives

$$\hat{p}_x = \hat{q}_s \frac{(A'Z_{a,t} + B')Z_{a,s}}{AZ_{a,t} + B + Z_{a,s}(CZ_{a,t} + D)}, \tag{7.47a}$$

$$\hat{q}_x = \hat{q}_s \frac{(C'Z_{a,t} + D')Z_{a,s}}{AZ_{a,t} + B + Z_{a,s}(CZ_{a,t} + D)}, \tag{7.47b}$$

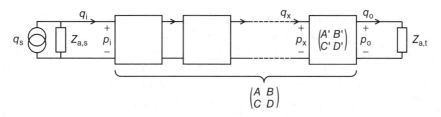

Figure 7.13 Calculation of the sound pressure inside a composite system

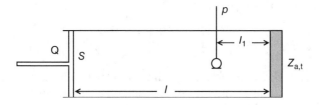

Figure 7.14 The sound pressure in a tube driven by a piston

Example 7.12 The sound pressure at an arbitrary position in a tube: The sound pressure at an arbitrary position in a tube driven by a piston with the volume velocity $Qe^{j\omega t}$ and terminated by the impedance $Z_{a,t}$ (see Figure 7.14) follows from Equation (7.47) with $Z_{a,s} = \infty$,

$$\hat{p} = \frac{A'Z_{a,t} + B'}{CZ_{a,t} + D}Qe^{j\omega t} = \frac{Z_{a,t}\cos kl_1 + j\dfrac{\rho c}{S}\sin kl_1}{j\dfrac{S}{\rho c}Z_{a,t}\sin kl + \cos kl}Qe^{j\omega t}.$$

Note that

$$\hat{p} = Qe^{j\omega t}\frac{\rho c}{S}e^{-jk(l-l_1)}$$

if $Z_{a,t} = \rho c/S$; in this case there is no reflection from the termination, so the tube acts as a simple delay. Note also that

$$\hat{p} = -jQe^{j\omega t}\frac{\rho c}{S}\frac{\cos kl_1}{\sin kl}$$

if the termination is rigid. At the piston (that is, for $l_1 = l$),

$$\hat{p} = -jQe^{j\omega t}\frac{\rho c}{S}\cot kl_1,$$

in agreement with Example 7.3.

7.3.2 System Performance

There are several ways of describing the performance of a silencer. The most useful measure is the *insertion loss*, which is defined as the ratio of the output mean-square pressure (or volume velocity) with a reference system to its value with the silencer (see Figure 7.15). The insertion loss is usually presented in logarithmic form. The reference system can for example be an infinitely short tube the transmission matrix of which is a unit matrix

$$\begin{pmatrix} A' & B' \\ C' & D' \end{pmatrix} = \begin{pmatrix} 1 & 0 \\ 0 & 1 \end{pmatrix} \tag{7.48}$$

(cf. Equation (7.43)), or a hard-walled tube that has the same length as the silencer. It follows from Equation (7.45) that

$$\text{IL} = 10\log\left(\frac{|AZ_{a,r} + B + Z_{a,s}(CZ_{a,r} + D)|^2}{|A'Z_{a,r} + B' + Z_{a,s}(C'Z_{a,r} + D')|^2}\right), \tag{7.49}$$

where A, B, C and D are the four-pole parameters of the entire system including the silencer and A', B', C' and D' are the corresponding parameters of the reference system. If this is an infinitely short tube Equation (7.49) reduces to

$$\text{IL} = 10\log\left(\frac{|AZ_{a,r} + B + Z_{a,s}(CZ_{a,r} + D)|^2}{|Z_{a,r} + Z_{a,s}|^2}\right). \tag{7.50}$$

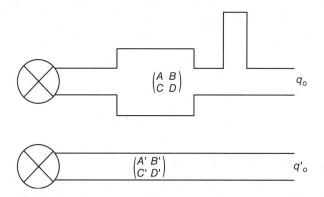

Figure 7.15 The insertion loss is obtained by comparing with a reference system

If the source is able to maintain its volume velocity irrespective of the acoustic load, then Equation (7.49) is reduced to

$$\text{IL} = 10\log\left(\frac{|CZ_{\text{a,r}} + D|^2}{|C'Z_{\text{a,r}} + D'|^2}\right), \tag{7.51}$$

and Equation (7.50) is simplified to

$$\text{IL} = 10\log(|CZ_{\text{a,r}} + D|^2). \tag{7.52}$$

It is not immediately apparent from Equation (7.49), but it can be shown that the insertion loss is independent of the length of the inlet pipe (the tube between the source and the silencer) if the source has an impedance of $\rho c/S$. The physical explanation is that a source with an impedance that equals the characteristic impedance of the connected tube does not reflect waves. In a similar manner it can be shown that the length of the tailpipe (the tube after the silencer) is of no importance if the terminating impedance is $\rho c/S$, rather than the radiation impedance $Z_{\text{a,r}}$ assumed in Equation (7.49). If both the source impedance and the termination impedance equal $\rho c/S$, then the length of the reference system is of no importance, since

$$\left|\frac{\rho c}{S}\cos kl + j\frac{\rho c}{S}\sin kl + \frac{\rho c}{S}\left(j\frac{S}{\rho c}\sin kl\frac{\rho c}{S} + \cos kl\right)\right|^2 = 4\left(\frac{\rho c}{S}\right)^2. \tag{7.53}$$

This particular version of the insertion loss with anechoic terminations at both ends is the *transmission loss*; Equation (7.49) now becomes

$$\text{TL} = 10\log\left(\frac{1}{4}\left|A + B + \frac{S}{\rho c} + C\frac{\rho c}{S} + D\right|^2\right). \tag{7.54}$$

It can be shown that the transmission loss is identical with the ratio of incident to transmitted sound power. Since the transmitted sound power cannot possibly exceed the incident sound power it follows that

$$\left|A + B\frac{S}{\rho c} + C\frac{\rho c}{S} + D\right| \geq 2 \tag{7.55}$$

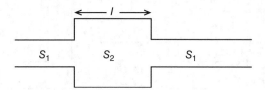

Figure 7.16   An expansion chamber

for *any* system.[2] By contrast, the insertion loss as defined by Equation (7.49) may well take negative values at certain frequencies (in dB), corresponding to the case where the performance of the silencer is poor compared with the performance of the reference system, for example because the input impedance of the reference system is particularly unfavourable.

Example 7.13 An expansion chamber: An expansion chamber is the simplest possible reactive silencer; see Figure 7.16. From Equations (7.43) and (7.54) it can be seen that the transmission loss is

$$\text{TL} = 10 \log \left(\frac{1}{4} \left| \cos kl + j \frac{S_1}{S_2} \sin kl + j \frac{S_2}{S_1} \sin kl + \cos kl \right|^2 \right)$$

$$= 10 \log \left(\cos^2 kl + \frac{1}{4} \left(h + \frac{1}{h} \right)^2 \sin^2 kl \right) = 10 \log \left(1 + \frac{1}{4} \left(h - \frac{1}{h} \right)^2 \sin^2 kl \right),$$

where h is the ratio of the cross-sectional areas and l is the length of the chamber; see Figure 7.17. Note that the transmission loss is zero when l is a multiple of half a wavelength.

Since the insertion loss takes account of reflections from the source and from the outlet of the tailpipe it gives a more realistic picture of the performance of the silencer than the transmission loss does. Figure 7.18 shows the insertion loss of an expansion chamber with an area expansion ratio of four, as in Figure 7.17, calculated for a constant volume velocity source. Note the significant influence of the reference system. With an infinitely short tube as the reference system the sharp peaks of the insertion loss disappear, because we no longer have particularly efficient radiation of *the reference system* at certain frequencies.

The impedance of the source can also have a significant influence on the insertion loss, as demonstrated by Figure 7.19, which shows the insertion loss of a tube with an infinitely

---

[2] Silencers of the reactive type contain no absorbing lining; they reduce the sound power radiation by impedance mismatch. Obviously, in the absence of dissipation in the silencer, the radiated sound power equals the sound power output of the source. The sound power *incident* on the silencer is, of course, larger than (or, at worst, equal to) the sound power output of the silencer.

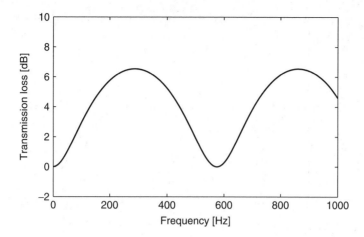

Figure 7.17   The transmission loss of an expansion chamber with an area expansion ratio of four

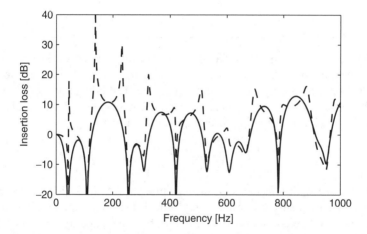

Figure 7.18   The insertion loss of an expansion chamber with an area expansion ratio of four, calculated under the assumption of a source of infinite internal impedance. Reference system: solid line, an infinitely short tube; dashed line, a tube of the same length

short tube as the reference system, calculated for a constant volume velocity source, for a constant sound pressure source, and for an anechoic source. The sound power output of the source is particularly large when it is loaded by an impedance comparable to the source impedance. Thus the sharp minima in the insertion loss occur when the input impedance of the tube 'matches' the impedance of the source. With a source of constant volume velocity this occurs when the effective length of the tube equals an odd-numbered multiple of a quarter of a wavelength; and with the constant sound pressure source the phenomenon occurs when the effective length of the tube equals a multiple of half a wavelength.

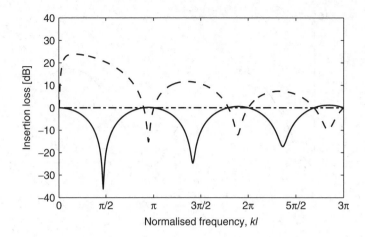

Figure 7.19 The insertion loss of a tube with hard walls, calculated for a source impedance of $100\ \rho c/S$ (solid line); $0.01\ \rho c/S$ (dashed line); $\rho c/S$ (dash-dotted line. The reference system is an infinitely short tube

Example 7.14 A side branch: The effect of a 'side branch resonator' (a closed tube mounted in the side of the system, cf. Figure 7.15) is to shunt the system with a low input impedance at certain frequencies. The boundary conditions imply continuity in the pressure but a sudden change in the volume velocity, cf. Equations (7.13) and (7.14), corresponding to the transmission matrix

$$\begin{pmatrix} A & B \\ C & D \end{pmatrix} = \begin{pmatrix} 1 & 0 \\ 1/Z_{a,i} & 1 \end{pmatrix}, \qquad (7.14\text{-}1)$$

where $Z_{a,i}$ is the input impedance of the side branch. A simple closed tube mounted in the side of an exhaust system is a very efficient silencer when the length of the tube is an odd-numbered multiple of a quarter of a wavelength (cf. Equation (7.10)). However, it has very little effect at other frequencies, as can be seen in Figure 7.20.

The closed, concentric tube that results if a pipe is projected into another wider pipe as shown in Figure 7.21 has the same effect as a closed tube mounted in the side of the system.

Example 7.15 A configuration with extended inlet and outlet: The relatively poor performance of an expansion chamber can be improved significantly if the inlet and outlet pipes are projected into the chamber, as shown in Figure 7.22. A particularly efficient configuration has recently been suggested by Chaitanaya and Munjal [4]; see Figure 7.23. It consists of an expansion chamber with projected inlet and outlet, and the good performance is due to the fact that the two resonators have their maximum effect at the frequencies where they cancel different troughs in the response of the chamber. However, good results require fine tuming of the lengths of the resonators. Chaitanaya and Munjal has provided numerically estimated expressions for end corrections for the extended inlet and outlet [4].

7.3.3 Dissipative Silencers

A dissipative silencer is a device that attenuates sound by absorbing acoustic energy, and a duct with porous lining materials on the walls can be an efficient silencer at medium frequencies. Such lined ducts are often used in air conditioning systems and also sometimes in combination with reactive silencers in exhaust systems.

There is no general solution in closed form to the problem of analysing sound propagation in a lined duct (see Section 7.7), and solving the problem usually involves iterative numerical procedures. The analysis leads to a solution in terms of waves with a complex wave number, where the imaginary part indicates that the waves decay exponentially as they propagate. Thus the transmission matrix of a lined duct of the length l and internal cross-sectional area S has the same form as Equation (7.43):

$$\begin{pmatrix} A & B \\ C & D \end{pmatrix} \simeq \begin{pmatrix} \cos((k - j\alpha)\,l) & j\dfrac{\rho c}{S}\sin((k - j\alpha)l) \\ j\dfrac{S}{\rho c}\sin((k - j\alpha)l) & \cos((k - j\alpha)l) \end{pmatrix}. \tag{7.56}$$

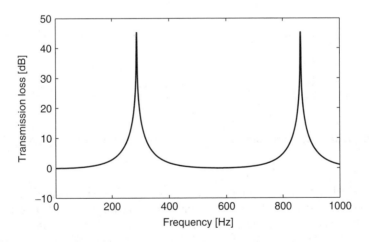

Figure 7.20 The transmission loss of a side branch resonator

Figure 7.21 Alternative side branch resonators: extended inlet and extended outlet

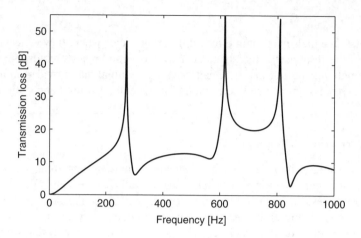

Figure 7.22   The transmission loss of an expansion chamber with extended inlet and outlet

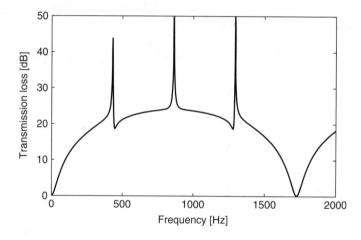

Figure 7.23   The transmission loss of a double-tuned extended inlet–outlet silencer

Dissipative silencers are in some respects easier to deal with than reactive silencers. If the lining attenuates sound waves so efficiently that the influence of sound reflected back from the tailpipe to the source can be neglected, then both the input and the output impedance of the silencer are approximately $\rho c / S$, from which it follows that the performance of the silencer is almost independent of the source and termination impedance, so that the insertion loss approximately equals the transmission loss. Moreover, the insertion loss (in dB) is simply proportional to the length of the duct, as can be seen as follows,

$$
\begin{aligned}
\mathrm{IL} \simeq \mathrm{TL} &\simeq 10 \log \left(|\cos \left((k - j\alpha) \, l \right) + j \sin((k - j\alpha)l)|^2 \right) \\
&= 20 \log \left(|e^{j(k-j\alpha)l}| \right) = 20 \log \left(e^{\alpha l} \right) = \frac{20}{\ln 10} \alpha l \simeq 8.69 \alpha l \ (\mathrm{dB}) \,.
\end{aligned}
\tag{7.57}
$$

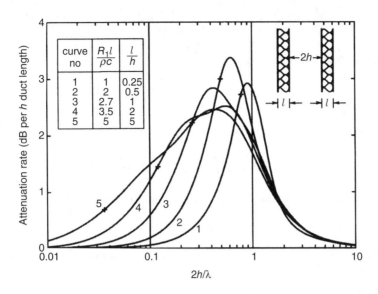

Figure 7.24 Attenuation rate of a rectangular duct lined on the two opposite sides with bulk reacting material (see Section 7.7.2). Lined circular ducts give twice the attenuation [8], reprinted by permission of Taylor & Francis

Curves showing the predicted attenuation of lined ducts of various configurations are given in many textbooks. An example is reproduced in Figure 7.24. The essential parameters are the flow resistivity of the lining material and the thickness of the lining relative to the duct width. A *very* brief account of the theory behind such predictions is given in Section 7.7.

7.4   Sound Propagation in Ducts with Mean Flow

The very purpose of the ducts of heating, ventilating and air-conditioning systems and engine exhaust pipes is to transport air, whereas the transmission of sound is an unwanted by-product. Whether mean flow affects the acoustic properties of such systems is therefore a matter of some practical importance.

In the general case where the flow is rotational and turbulent the analysis is very complicated. In what follows we shall only touch upon the simple case of ducts with steady, uniform flow. Equations (2.12) and (2.13) now become

$$\nabla \cdot \mathbf{u} + \frac{1}{\rho c^2} \frac{\partial p}{\partial t} + \frac{1}{\rho c^2} \mathbf{u_0} \cdot \nabla p = 0, \tag{7.58}$$

$$\nabla p + \rho \frac{\partial \mathbf{u}}{\partial t} + \rho (\mathbf{u_0} \cdot \nabla)\mathbf{u} = \mathbf{0}, \tag{7.59}$$

where $\mathbf{u_0}$ is the mean flow velocity. If we assume that the flow is axial (in the z-direction) and the sound field is one-dimensional, Equations (7.58) and (7.59) become

$$\frac{\partial u_z}{\partial z} + \frac{1}{\rho c^2} \frac{\partial p}{\partial t} + \frac{1}{\rho c^2} u_0 \frac{\partial p}{\partial z} = 0, \tag{7.60}$$

$$\frac{\partial p}{\partial z} + \rho \frac{\partial u_z}{\partial t} + \rho u_0 \frac{\partial u_z}{\partial z} = 0. \tag{7.61}$$

Eliminating u_z from these two equations gives a modified wave equation:

$$\frac{\partial^2 p}{\partial z^2}\left(1 - \left(\frac{u_0}{c}\right)^2\right) - \frac{1}{c^2}\frac{\partial^2 p}{\partial t^2} - \frac{2u_0}{c^2}\frac{\partial^2 p}{\partial t \partial z} = 0. \tag{7.62}$$

The corresponding equation in the usual complex notation is

$$\frac{\partial^2 \hat{p}}{\partial z^2}(1 - M^2) - k^2 \hat{p} - 2jkM \frac{\partial \hat{p}}{\partial z} = 0, \tag{7.63}$$

where we have also introduced the *Mach number* of the flow,

$$M = u_0/c \tag{7.64}$$

It is fairly easy to show that the general solution to Equation (7.63) can be written in the form

$$\hat{p} = p_+ e^{j(\omega t - k_+ l)} + p_- e^{j(\omega t + k_- l)}, \tag{7.65}$$

where

$$k_+ = \frac{k}{1 + M}, \text{ and } k_- = \frac{k}{1 - M}. \tag{7.66a, 7.66b}$$

Since $k = 2\pi/\lambda$ it can be seen that the effect of the flow is to stretch the wavelength of waves travelling in the direction of the flow by a factor of $1 + M$, corresponding to an increase of the speed of sound,

$$c_+ = c(1 + M) = c + u_0, \tag{7.67a}$$

and to contract the wavelength of waves travelling in the opposite direction by a factor of $1 - M$, corresponding to a decrease of the speed of sound,

$$c_- = c(1 - M) = c - u_0, \tag{7.67b}$$

which of course is just what one would expect.

The transmission matrix of a tube with the length l, the cross-sectional area S and mean flow in the z-direction with a Mach number of M can now be calculated in the same manner as used in Section 7.3.1. The result is

$$\begin{pmatrix} A & B \\ C & D \end{pmatrix} = e^{-jMk_e l} \begin{pmatrix} \cos k_e l & j\frac{\rho c}{S} \sin k_e l \\ j\frac{S}{\rho c} \sin k_e l & \cos k_e l \end{pmatrix}, \tag{7.68}$$

where

$$k_e = \frac{k}{1 - M^2}. \tag{7.69}$$

Since $k_e > k$ it follows that the flow tends to reduce resonance and antiresonance frequencies. However, with the moderate air speeds found in heating, ventilating and air-conditioning systems and in engine exhausts the effect is negligible. Since there is no flow in a side branch resonator the effect of even a slight modification of the wavelength in some of the pipes of a composite system can be significant, though. For example, the performance of the double-tuned system shown in Figure 7.23 is more affected by mean flow than the performance of the simple expansion chamber.

It should finally be mentioned that mean flow also affects the radiation of sound power from the tailpipe. It can be shown [6] that the sound intensity is given by

$$I_z = (1 + M^2)\frac{1}{2}\text{Re}\{\hat{p}\hat{u}_z^*\} + M\left(\frac{|\hat{p}|^2}{2\rho c} + \frac{\rho c}{2}|\hat{u}_z|^2\right).$$ (7.70)

7.5 Three-dimensional Waves in Ducts with Rigid Walls

The foregoing considerations have been based on the assumption that the sound field in a uniform duct is one-dimensional. As we shall see this is a good approximation at low frequencies, but at high frequencies the sound field in a duct is more complicated. It is possible to solve the problem analytically only for uniform ducts with a simple cross-sectional shape. The analytical method is classical and known as 'separation of variables'.

7.5.1 The Sound Field in a Duct with Rectangular Cross Section

In the usual rectangular coordinate system the Helmholtz equation takes the form

$$\frac{\partial^2 \hat{p}}{\partial x^2} + \frac{\partial^2 \hat{p}}{\partial y^2} + \frac{\partial^2 \hat{p}}{\partial z^2} + k^2 \hat{p} = 0.$$ (7.71)

The solution to this equation must also satisfy the boundary conditions implied by the rigid walls. With the coordinate system sketched in Figure 7.25 the boundary conditions are

$$\hat{u}_x(0, y, z) = 0, \quad \hat{u}_x(a, y, z) = 0,$$ (7.72a, 7.72b)

and

$$\hat{u}_y(x, 0, z) = 0, \quad \hat{u}_y(x, b, z) = 0,$$ (7.73a, 7.73b)

which implies that

$$\frac{\partial \hat{p}}{\partial x} = 0$$ (7.74)

at $x = 0$ and $x = a$, and that

$$\frac{\partial \hat{p}}{\partial y} = 0$$ (7.75)

at $y = 0$ and $y = b$.

It is now assumed that the solution to Equation (7.71) can be written as a product of three factors each of which depends on only one coordinate, which means that the sound pressure can be written as

$$\hat{p}(x, y, z, t) = p_x(x)p_y(y)p_z(z)e^{j\omega t}.$$ (7.76)

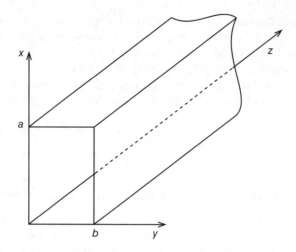

Figure 7.25   A tube with rectangular cross section

Insertion into Equation (7.71) gives

$$p_y(y)p_z(z)\frac{d^2 p_x(x)}{dx^2} + p_x(x)p_z(z)\frac{d^2 p_y(y)}{dy^2} + p_x(x)p_y(y)\frac{d^2 p_z(z)}{dz^2}$$

$$+k^2 p_x(x)p_y(y)p_z(z) = 0 \qquad (7.77)$$

or

$$\frac{1}{p_x(x)}\frac{d^2 p_x(x)}{dx^2} + \frac{1}{p_y(y)}\frac{d^2 p_y(y)}{dy^2} + \frac{1}{p_z(z)}\frac{d^2 p_z(z)}{dz^2} + k^2 = 0. \qquad (7.78)$$

Inspection of Equation (7.78) leads to the conclusion that the three first terms must be independent of x, y and z. For example, the first term is obviously independent of y and z, and it does not depend on x either, since it equals the sum of three terms that do not depend on x. Equating this term with the constant $-k_x^2$ gives

$$\frac{d^2 p_x(x)}{dx^2} + k_x^2 p_x(x) = 0, \qquad (7.79)$$

which is recognised as the one-dimensional Helmholtz equation with the general solution

$$p_x(x) = Ae^{-jk_x x} + Be^{jk_x x}, \qquad (7.80)$$

(cf. Equation (3.11)). In a similar manner it follows that

$$p_y(y) = Ce^{-jk_y y} + De^{jk_y y} \qquad (7.81)$$

and

$$p_z(z) = Ee^{-jk_z z} + Fe^{jk_z z}. \qquad (7.82)$$

Consequently, the general solution to Equation (7.71) can be written

$$\hat{p} = (Ae^{-jk_x x} + Be^{jk_x x})(Ce^{-jk_y y} + De^{jk_y y})(Ee^{-jk_z z} + Fe^{jk_z z})e^{j\omega t}, \qquad (7.83)$$

where the constants k_x, k_y and k_z must satisfy the condition

$$k_x^2 + k_y^2 + k_z^2 = k^2. \tag{7.84}$$

The boundary condition at $x = 0$ (Equation (7.74)) implies that

$$-jk_x A e^{-jk_x x} + jk_x B e^{jk_x x} = 0 \tag{7.85}$$

at $x = 0$, from which it follows that $A = B$. Accordingly,

$$\frac{dp_x(x)}{dx} = -2k_x A \sin k_x x, \tag{7.86}$$

which shows that the boundary condition at $x = a$ can be satisfied only for certain discrete values of k_x, determined by the condition

$$k_x a = m\pi \tag{7.87}$$

where m is an integer. Obviously, the boundary conditions at $y = 0$ and $y = b$ lead to a similar condition for k_y,

$$k_y b = n\pi \tag{7.88}$$

where n is an integer. From Equation (7.84) it follows that

$$\left(\frac{m\pi}{a}\right)^2 + \left(\frac{n\pi}{b}\right)^2 + k_z^2 = k^2, \tag{7.89}$$

which shows that the axial wavenumber k_z depends on m and n. Equation (7.83) now becomes

$$\hat{p} = \sum_{m=0}^{\infty} \sum_{n=0}^{\infty} \sqrt{\varepsilon_m \varepsilon_n} \left(\cos\left(\frac{m\pi x}{a}\right) \cos\left(\frac{n\pi y}{b}\right)\right) \left(p_{mn+} e^{j(\omega t - k_{zmn} z)} + p_{mn-} e^{j(\omega t + k_{zmn} z)}\right), \tag{7.90}$$

where

$$k_{zmn} = \left(k^2 - \left(\frac{m\pi}{a}\right)^2 - \left(\frac{n\pi}{b}\right)^2\right)^{1/2}, \tag{7.91}$$

and $\varepsilon_m = 1$ for $m = 0$ and $\varepsilon_m = 2$ for $m > 0$. (The factor $\sqrt{\varepsilon_m \varepsilon_n}$ is a normalisation constant; see Section 7.6.) Each term in the sum represents a wave or a 'duct mode' and a corresponding reflected mode. The term in the first parenthesis is the corresponding cross-sectional eigenfunction. With $m = n = 0$,

$$\hat{p} = p_+ e^{j(\omega t - kz)} + p_- e^{j(\omega t + kz)}, \tag{7.92}$$

which can be recognised as the one-dimensional sound field assumed in Sections 7.2 and 7.3. The plane wave is the fundamental duct mode.

The general term with $(m, n) \neq (0, 0)$ is a higher-order mode with the propagation wavenumber k_{zmn}. This mode can propagate unattenuated in the positive or negative z-direction if k_{zmn} is real, which is the case if

$$k = \frac{\omega}{c} \geq \left(\left(\frac{m\pi}{a}\right)^2 + \left(\frac{n\pi}{b}\right)^2\right)^{1/2}. \tag{7.93}$$

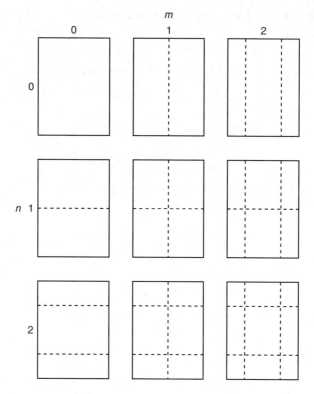

Figure 7.26 Nodal lines of (m, n)-modes in a duct with rectangular cross section

Inspection of Equation (7.90) shows that the sound pressure of the (m, n)-mode is identically zero in m 'nodal planes' parallel to the yz-plane and in n nodal planes parallel to the xz-plane; cf. Figure 7.26.

The axial particle velocity component associated with the (m, n)-mode follows from Euler's equation of motion and Equation (7.90),

$$\hat{u}_{zmn} = \frac{P_{mn+}}{\rho c} \sqrt{\varepsilon_m \varepsilon_n} \frac{k_{zmn}}{k} \cos\left(\frac{m\pi x}{a}\right) \cos\left(\frac{n\pi y}{b}\right) e^{j(\omega t - k_{zmn} z)}, \qquad (7.94)$$

where, for simplicity, we have ignored the reflected wave. The corresponding axial component of the sound intensity associated with the mode is

$$
\begin{aligned}
I_{zmn}(x, y, z) &= \frac{1}{2}\mathrm{Re}\{\hat{p}\hat{u}_{zmn}^*\} \\
&= \frac{|P_{mn+}|^2}{2\rho c}\varepsilon_m \varepsilon_n \cos^2\left(\frac{m\pi x}{a}\right)\cos^2\left(\frac{n\pi y}{b}\right)\mathrm{Re}\{k_{zmn}/k\}, \qquad (7.95)
\end{aligned}
$$

which is seen to be independent of z, as expected. The particle velocity component in the x-direction, is

$$\hat{u}_{xmn} = -j\frac{P_{mn+}}{\rho c}\sqrt{\varepsilon_m \varepsilon_n}\frac{m\pi}{ka}\sin\left(\frac{m\pi x}{a}\right)\cos\left(\frac{n\pi y}{b}\right)e^{j(\omega t - k_{zmn} z)}, \qquad (7.96)$$

which clearly is in quadrature with the sound pressure; therefore the sound intensity in a single isolated (m, n)-mode is zero in the transversal direction. In the general case the local sound intensity distribution is far more complicated, because the sound pressure in any mode cooperates with the particle velocity in any mode. However, as we shall see later, since the eigenfunctions are orthogonal such cross terms integrate to zero over the cross section of the duct, which means that the total sound power transmitted through the duct is simply the sum of the powers associated with the various modes.

The expression for an (m, n)-mode that propagates along the tube can be rewritten as follows,

$$
\begin{aligned}
\hat{p}_{mn} &= \frac{p_{mn+}}{4} \sqrt{\varepsilon_m \varepsilon_n} (e^{-jk_x x} + e^{jk_x x})(e^{-jk_y y} + e^{jk_y y}) e^{j(\omega t - k_z z)} \\
&= \frac{p_{mn+}}{4} \sqrt{\varepsilon_m \varepsilon_n} \left(e^{j(\omega t - k_x x - k_y y - k_z z)} + e^{j(\omega t - k_x x + k_y y - k_z z)} + e^{j(\omega t + k_x x + k_y y - k_z z)} \right. \\
&\qquad \left. + e^{j(\omega t + k_x x - k_y y - k_z z)} \right)
\end{aligned}
\tag{7.97}
$$

(cf. Equation (7.83)), which shows that the mode can be decomposed into four interfering plane waves propagating in the directions (k_x, k_y, k_z), $(k_x, -k_y, k_z)$, $(-k_x, k_y, k_z)$, and $(-k_x, -k_y, k_z)$. These plane waves may be regarded as one obliquely incident plane wave and its reflections from the walls of the tube. The waves interfere in such a manner that the boundary conditions (Equations (7.74) and (7.75)) are satisfied.

Since $k_z < k$, the *apparent wavelength* in the z-direction is longer than the wavelength in free space, λ $(= c/f)$:

$$
\lambda_z = 2\pi/k_z = \frac{\lambda}{\sqrt{1 - (k_x/k)^2 - (k_y/k)^2}}.
\tag{7.98}
$$

Accordingly, the *phase velocity* exceeds the speed of sound:

$$
c_p = \lambda_z f = \frac{\omega}{k_z} = \frac{c}{\sqrt{1 - (k_x/k)^2 - (k_y/k)^2}}.
\tag{7.99}
$$

The phase velocity is the velocity with which a given phase angle (or wave front) moves along in the tube. On the other hand, the fact that the four interfering plane waves that constitute the mode travel at oblique angles relative to the z-direction indicates that the sound energy propagates at a velocity that is *lower* than c. It can be shown that this speed, known as the *group velocity*, is:

$$
c_g = \left(\frac{\partial k_z}{\partial \omega} \right)^{-1} = c\sqrt{1 - (k_x/k)^2 - (k_y/k)^2}.
\tag{7.100}
$$

This is the velocity with which a wavelet composed of a narrow band of frequencies (or the envelope of a similar noise signal) travels along the tube [7]. Note that c_g depends on the frequency; this phenomenon, which is known as *dispersion*, means that the various frequency components of a wide-band signal travel with different velocities. A broad band impulse will change its shape as it travels along the tube.

In the foregoing we have assumed that the condition expressed by the inequality (7.93) was satisfied. If, by contrast, the tube is driven with the frequency determined by the condition

$$k = \left(\left(\frac{m\pi}{a} \right)^2 + \left(\frac{n\pi}{b} \right)^2 \right)^{1/2},$$

(7.101)

that is, at the frequency

$$f_{mn} = \frac{c}{2} \left(\left(\frac{m}{a} \right)^2 + \left(\frac{n}{b} \right)^2 \right)^{1/2},$$

(7.102)

then $k_{zmn} = 0$, and the sound pressure associated with the (m, n)-mode can be written

$$\hat{p}(x, y, z, t) = p_{mn+} \sqrt{\varepsilon_m \varepsilon_n} \cos \left(\frac{m\pi x}{a} \right) \cos \left(\frac{n\pi y}{b} \right) e^{j\omega t}.$$

(7.103)

In this case we have a two-dimensional transversal standing wave that does not depend on z at all. At a still lower frequency, when

$$k < \left(\left(\frac{m\pi}{a} \right)^2 + \left(\frac{n\pi}{b} \right)^2 \right)^{1/2},$$

(7.104)

the propagation wavenumber k_{zmn} becomes purely imaginary. The corresponding mode is attenuated exponentially,

$$\hat{p}_{mn}(x, y, z, t) = p_{mn+} \sqrt{\varepsilon_m \varepsilon_n} \cos \left(\frac{m\pi x}{a} \right) \cos \left(\frac{n\pi y}{b} \right) e^{-\gamma_z z} e^{j\omega t},$$

(7.105)

where

$$\gamma_z = \left(\left(\frac{m\pi}{a} \right)^2 + \left(\frac{n\pi}{b} \right)^2 - k^2 \right)^{1/2} = \frac{2\pi}{c} \left(f_{mn}^2 - f^2 \right)^{1/2}$$

(7.106)

is the (positive and real-valued) propagation coefficient. The axial component of the particle velocity is

$$\hat{u}_{zmn}(x, y, z, t) = -j \frac{p_{mn+}}{\rho c} \sqrt{\varepsilon_m \varepsilon_n} \frac{\gamma_z}{k} \cos \left(\frac{m\pi x}{a} \right) \cos \left(\frac{n\pi y}{b} \right) e^{-\gamma_z z} e^{j\omega t},$$

(7.107)

which is seen to be 90° out of phase with the pressure; therefore the sound intensity is zero (this can also be seen from Equation (7.95) when k_{zmn} is imaginary). Such a mode is called an *evanescent mode*.

The foregoing considerations can be summarised as follows. In the general case the sound field in the tube can be decomposed into a sum of modes or waves, the plane wave plus an infinite number of waves of higher order. Each higher-order mode has a cutoff frequency determined by Equation (7.102); if driven below this frequency the mode is evanescent and decays exponentially with the distance; if driven above this frequency the mode propagates unattenuated. If the tube is driven with a frequency that is lower than the lowest cutoff frequency (which corresponds to the width of the tube being half a wavelength), only the fundamental mode (the plane wave) can propagate, and all other modes decrease exponentially. This is the justification of the simplifying assumption made in Sections 7.2, 7.3 and 7.4.

The cutoff frequency of the lowest higher-order mode (1, 0) is the frequency at which a equals half a wavelength (if $a > b$).

Example 7.16 The decay of an evanescent mode: A given (m, n)-mode is driven an octave below its cutoff frequency, that is, at the frequency $f = f_{mn}/2$. From Equation (7.106) it can be seen that

$$\gamma_z = \frac{2\pi}{c}\left(f_{mn}^2 - (f_{mn}/2)^2\right)^{1/2} = \frac{\pi f_{mn}}{c}\sqrt{3} = \frac{2\pi f}{c}\sqrt{3}.$$

The mode decays rapidly with the distance. For example, the attenuation over a distance that corresponds to the free-field wavelength at the given frequency is

$$\Gamma = 20\log\left(e^{\gamma_z\lambda}\right) = 20\log\left(\exp\left(2\pi\sqrt{3}\right)\right) \simeq 94.5\,\text{dB}.$$

Example 7.17 Microphone arrangements for separating modes: Since the cross-sectional eigenfunctions are known and simple it is possible to detect various modes in a rectangular duct with an appropriate arrangement of microphones. For example, an arrangement consisting of two pressure microphones placed symmetrically about the nodal plane of the (1, 0)-mode (say, at $(a/4, b/2, z_0)$ and $(3a/4, b/2, z_0)$ as sketched in Figure 7.27) makes it possible to separate the plane wave from the (1, 0)-mode in the frequency range where only these two modes can propagate. Clearly, with

$$\hat{p}(x, y, z, t) = \left(Ae^{-jkz} + Be^{jkz} + C\cos\left(\frac{\pi x}{a}\right)e^{-jk'z} + D\cos\left(\frac{\pi x}{a}\right)e^{jk'z}\right)e^{j\omega t},$$

the sum of the microphone signals is proportional to the one-dimensional part of the sound field,

$$\hat{p}(a/4, b/2, z_0) + \hat{p}(3a/4, b/2, z_0) = 2(Ae^{-jkz_0} + Be^{jkz_0})e^{j\omega t},$$

whereas the difference is proportional to the (1, 0)-mode,

$$\hat{p}(a/4, b/2, z_0) - \hat{p}(3a/4, b/2, z_0) = \sqrt{2}(Ce^{-jk'z_0} + De^{jk'z_0})e^{j\omega t}.$$

Placing the microphones in the xz-plane at $y = b/2$ has the additional advantage that the (0, 1)-mode has no influence on the detection.

7.5.2 The Sound Field in a Duct with Circular Cross Section

Most ducts have a circular cross section. In order to determine the sound field in such a duct one must express the Helmholtz equation in a cylindrical coordinate system

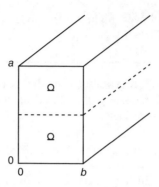

Figure 7.27   Microphone arrangement for separating modes (0, 0) and (1, 0)

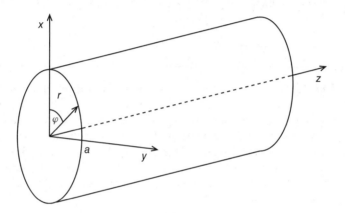

Figure 7.28   A tube with circular cross section

(see Figure 7.28):

$$\frac{\partial^2 \hat{p}}{\partial r^2} + \frac{1}{r}\frac{\partial \hat{p}}{\partial r} + \frac{1}{r^2}\frac{\partial^2 \hat{p}}{\partial \varphi^2} + \frac{\partial^2 \hat{p}}{\partial z^2} + k^2 \hat{p} = 0, \tag{7.108}$$

cf. Equation (2.20b). The boundary condition for a tube with rigid walls is

$$\left.\frac{\partial \hat{p}}{\partial r}\right|_{r=a} = 0, \tag{7.109}$$

It is now assumed that the sound pressure can be expressed as the product of functions that depend only on one coordinate:

$$\hat{p}(r, \varphi, z, t) = p_r(r)p_\varphi(\varphi)p_z(z)e^{j\omega t}. \tag{7.110}$$

Insertion gives

$$\frac{1}{p_r(r)}\frac{\mathrm{d}^2 p_r(r)}{\mathrm{d}r^2} + \frac{1}{r}\frac{1}{p_r(r)}\frac{\mathrm{d}p_r(r)}{\mathrm{d}r} + \frac{1}{r^2}\frac{1}{p_\varphi(\varphi)}\frac{\mathrm{d}^2 p_\varphi(\varphi)}{\mathrm{d}\varphi^2} + \frac{1}{p_z(z)}\frac{\mathrm{d}^2 p_z(z)}{\mathrm{d}z^2} + k^2 = 0. \tag{7.111}$$

It is apparent that the last term but one on the left-hand side is a function only of z that equals a sum of various terms that are independent of z; therefore it must be a constant, $-k_z^2$

$$\frac{d^2 p_z(z)}{dz^2} + k_z^2 p_z(z) = 0. \tag{7.112}$$

Inserting and multiplying with r^2 gives

$$\frac{r^2}{p_r(r)} \frac{d^2 p_r(r)}{dr^2} + \frac{r}{p_r(r)} \frac{dp_r(r)}{dr} + \frac{1}{p_\varphi(\varphi)} \frac{d^2 p_\varphi(\varphi)}{d\varphi^2} + r^2(k^2 - k_z^2) = 0. \tag{7.113}$$

A similar argumentation now shows that

$$\frac{d^2 p_\varphi(\varphi)}{d\varphi^2} + k_\varphi^2 p_\varphi(\varphi) = 0, \tag{7.114}$$

where $-k_\varphi^2$ is the third term of Equation (7.113), and therefore

$$\frac{d^2 p_r(r)}{dr^2} + \frac{1}{r} \frac{dp_r(r)}{dr} + p_r(r) \left(k^2 - k_z^2 - \frac{k_\varphi^2}{r^2} \right) - 0. \tag{7.115}$$

Equations (7.112) and (7.114) can be identified as versions of the one-dimensional Helmholtz equation with the general solutions

$$p_z(z) = A e^{-jk_z z} + B e^{jk_z z} \tag{7.116}$$

and

$$p_\varphi(\varphi) = C e^{-jk_\varphi \varphi} + D e^{jk_\varphi \varphi} \tag{7.117}$$

(cf. Equation (3.11)). Inspection of Equation (7.117) leads to the conclusion that k_φ must be an integer, since the function p_φ should obviously be periodic with period 2π:

$$p_\varphi(\varphi) = p_\varphi(\varphi + 2\pi). \tag{7.118}$$

Accordingly, we shall use the symbol m instead of k_φ for this integer. Moreover, often we can combine the two exponentials in Equation (7.117),

$$p_\varphi(\varphi) = E \cos(m\varphi + \varphi_m). \tag{7.119}$$

(This is expanded later). Finally, we introduce the transversal wavenumber k_r:

$$k_r^2 = k^2 - k_z^2. \tag{7.120}$$

Equation (7.115) now becomes

$$\frac{d^2 p_r(r)}{dr^2} + \frac{1}{r} \frac{dp_r(r)}{dr} + p_r(r) \left(k_r^2 - \frac{m^2}{r^2} \right) = 0. \tag{7.121}$$

This differential equation is known as Bessel's equation, with the general solution

$$p_r(r) = F J_m(k_r r) + G N_m(k_r r), \tag{7.122}$$

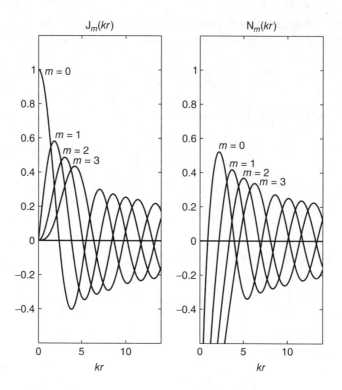

Figure 7.29 Bessel and Neumann functions of order m for $m = 0, 1, \ldots, 3$

where J_m is a Bessel function of order m and N_m is a Neumann function of order m. (Neumann functions are also known as Bessel functions of the second kind.) See Figure 7.29 and Appendix C. Only the Bessel functions are finite at $r = 0$; therefore G must be zero.

From the boundary condition, Equation (7.109), it follows that

$$\frac{dJ_m(k_r r)}{dr}\bigg|_{r=a} = J'_m(k_r a) = 0. \tag{7.123}$$

This equation can be satisfied only for certain discrete values of k_r, denoted k_{rmn} with $n = 1, 2, 3, \ldots$ (see Figure 7.30 and Table 7.1). The solution to Equations (7.108) and (7.109) can now be written

$$\hat{p}(r, \varphi, z, t) = \sum_{m=-\infty}^{\infty} \sum_{n=0}^{\infty} \Lambda_{mn} \sqrt{\varepsilon_m} J_m(k_{rmn} r) \cos(m\varphi + \varphi_m)$$

$$(p_{mn+} e^{j(\omega t - k_{zmn} z)} + p_{mn-} e^{j(\omega t + k_{zmn} z)}), \tag{7.124}$$

where

$$k_{zmn} = (k^2 - k_{rmn}^2)^{1/2}, \tag{7.125}$$

and

$$\Lambda_{mn} = (1 - m^2/(k_{rmn} a)^2)^{-1/2}/J_m(k_{rmn} a) \tag{7.126}$$

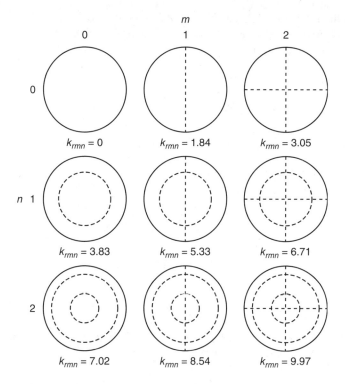

Figure 7.30 Nodal lines of (m, n)-modes in a duct with circular cross section. The orientation of the nodal planes that occur when $m > 0$ depends on φ_m

Table 7.1 Roots of Equation (7.123)

$n \setminus m$	0	1	2	3
0	0	1.8412	3.0542	4.2012
1	3.8317	5.3314	6.7061	8.0152
2	7.0156	8.5363	9.9695	11.3459
3	10.1735	11.7060	13.1704	14.5859

is a normalisation constant. Each term in the sum represents a mode. The modes corresponding to $m = 0$ are axisymmetric. With $m = n = 0$ Equation (7.124) simplifies to

$$\hat{p}(r, \varphi, z, t) = p_+ e^{j(\omega t - kz)} + p_- e^{j(\omega t + kz)}, \tag{7.127}$$

which is recognised as the one-dimensional sound field dealt with in Sections 7.2 and 7.3. In the general case there are m nodal planes (with an orientation that depends on φ_m) where the sound pressure is zero, and n coaxial nodal cylindrical surfaces; see Figure 7.30. Each mode has a cutoff frequency that follows from Equation (7.123) with $k_{rmn} = 0$,

$$f_{mn} = \frac{\eta_{mn} c}{2\pi a}, \tag{7.128}$$

where η_{mn} is the n'th root of Equation (7.123). Some of these roots are listed in Table 7.1. If it is driven above the frequency given by Equation (7.128) the mode can propagate unattenuated with the phase velocity

$$c_p = \frac{\omega}{k_{zmn}} = \frac{c}{\sqrt{1-(k_{rmn}/k)^2}} = \frac{c}{\sqrt{1-(f_{mn}/f)^2}}, \tag{7.129}$$

cf. Equation (7.99), whereas the group velocity is

$$c_g = \left(\frac{\partial k_z}{\partial \omega}\right)^{-1} = c\sqrt{1-(k_{rmn}/k)^2} = c\sqrt{1-(f_{mn}/f)^2}, \tag{7.130}$$

cf. Equation (7.100). If it is driven below the cutoff frequency the mode declines exponentially with the propagation coefficient

$$\gamma_z = \sqrt{k_{rmn}^2 - k^2} = \frac{2\pi}{c}\sqrt{f_{mn}^2 - f^2}, \tag{7.131}$$

cf. Equation (7.106).

It can be concluded that the first axisymmetric higher-order mode (0, 1) can propagate when $ka > 3.83$, corresponding to the condition

$$f > f_{01} \simeq \frac{3.83c}{2\pi a}. \tag{7.132}$$

However, the lowest cutoff frequency in a duct of circular cross section is the cutoff frequency of the first non-axisymmetric mode (1, 0),

$$f_{10} \simeq \frac{1.84c}{2\pi a}; \tag{7.133}$$

below this frequency only the plane wave can propagate.

Example 7.18 A cylindrical duct with a central kernel: The sound field in a duct consisting of the space between two concentric rigid cylinders with radii a and b (see Figure 7.31) can be written as a sum of modes of the form

$$\hat{p}_m(r, \varphi, z, t) = (AJ_m(k_{rm}r) + BN_m(k_{rm}r))e^{j(\omega t - m\varphi - k_{zm}z)},$$

since we cannot discard the 'Neumann term' in this case. The ratio of A to B follows from the boundary condition at $r = a$,

$$AJ'_m(k_{rm}a) + BN'_m(k_{rm}a) = 0,$$

from which it can be seen that

$$\frac{A}{B} = -\frac{N'_m(k_{rm}a)}{J'_m(k_{rm}a)}.$$

The boundary condition at $r = b$ implies that

$$\frac{A}{B} = -\frac{N'_m(k_{rm}b)}{J'_m(k_{rm}b)},$$

which leads to the conclusion that both boundary conditions can be satisfied only for certain discrete values of k_{rm} (denoted k_{rmn}), determined by the condition

$$N'_m(k_{rmn}a)J'_m(k_{rmn}b) = N'_m(k_{rmn}b)J'_m(k_{rmn}a).$$

In the special case where the ratio of a to b equals the ratio of two of the roots of Equation (7.123), say (with $m = 1$),

$$\frac{a}{b} = \frac{5.331}{1.841},$$

the above condition is fulfilled for

$$k_{rmn} = \frac{5.331}{a} = \frac{1.841}{b}.$$

In this particular case there is no Neumann term, that is, $B = 0$; the 'Bessel term' satisfies the boundary conditions both at $r = a$ and at $r = b$.

Equation (7.119) is actually a special case, not the general solution to Equation (7.114). Let us ignore the reflected waves for a moment. The general term,

$$\hat{p}_{mn}(r, \varphi, z, t) = \tilde{p}_{mn+}\Lambda_{mn}J_m(k_{rmn}r)e^{j(\omega t - m\varphi - k_{zmn}z)}, \tag{7.134}$$

is a *spinning mode* or helical wave (for $m = 0$) with the circumferential mode number m (a positive or negative integer) and the radial mode number n. There are no nodal planes in a spinning mode; the wavefronts are spirals, hence the name; see Figure 7.32.[3] Spinning modes are generated by rotating sources as, e.g., fans and propellers. However, many other excitation mechanisms tend to generate spinning modes in pairs with the same amplitude, and we can now reinterpret Equation (7.124) as the sum of such pairs

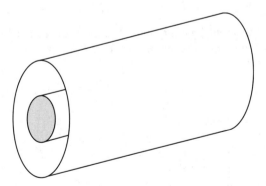

Figure 7.31 A circular duct with a central, coaxial, solid cylinder

---

[3] The eigenfunctions in a duct are usually, but not always, real functions. Spinning modes in ducts with circular cross section correspond to complex eigenfunctions.

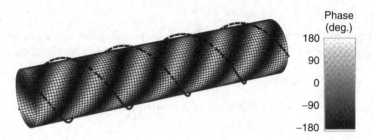

Figure 7.32   Wavefronts of a spinning mode [8]

of interfering spinning modes (with m and $-m$), spiralling in opposite directions with the same amplitude.

Example 7.19 A partitioned duct: It is easy to see that modes of the form

$$\hat{p}_{mn}(r, \varphi, z, t) = \Lambda_{mn}\sqrt{\varepsilon_m}\,\mathrm{J}_m(k_{rmn}r)\cos(m\varphi)\mathrm{e}^{\mathrm{j}(\omega t - k_{zmn}z)}$$

can propagate in the partitioned duct shown in Figure 7.33. A thin rigid partition gives rise to the additional boundary condition

$$\frac{\partial \hat{p}_{mn}(r, \varphi, z, t)}{\partial \varphi} = 0 \qquad \text{for} \qquad \varphi = 0 \qquad \text{and} \qquad \varphi = \pi,$$

and this is obviously satisfied by the above expression. The partition excludes spinning modes and fixes the position of the nodal lines determined by the quantity φ_m in Equation (7.119).

7.5.3   The Sound Field in a Duct with Arbitrary Cross-sectional Shape

We shall now study a duct with arbitrary but constant cross-sectional shape. In the general case one cannot solve the problem analytically. However, the separation method can still

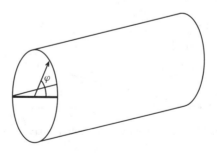

Figure 7.33   A partitioned duct with circular cross section

be used. On the assumption that the sound pressure can be written as the product

$$\hat{p}(x, y, z, t) = p_{xy}(x, y) p_z(z) e^{j\omega t}, \tag{7.135}$$

the Helmholtz equation becomes

$$p_z(z) \left(\frac{\partial^2}{\partial x^2} + \frac{\partial^2}{\partial y^2} \right) p_{xy}(x, y) + p_{xy}(x, y) \frac{\partial^2 p_z(z)}{\partial z^2} + k^2 p_{xy}(x, y) p_z(z) = 0, \tag{7.136}$$

which shows that

$$\frac{1}{p_{xy}(x, y)} \left(\frac{\partial^2}{\partial x^2} + \frac{\partial^2}{\partial y^2} \right) p_{xy}(x, y) + \frac{1}{p_z(z)} \frac{d^2 p_z(z)}{dz^2} + k^2 = 0. \tag{7.137}$$

Irrespective of the cross-sectional shape, the two-dimensional Helmholtz equation

$$\left(\frac{\partial^2}{\partial x^2} + \frac{\partial^2}{\partial y^2} \right) p_{xy}(x, y) + k^2 p_{xy}(x, y) = 0 \tag{7.138}$$

with the boundary condition that the normal component of the gradient is zero at the walls,

$$\nabla p_{xy}(x, y) \cdot \mathbf{n}_{\text{wall}} = 0, \tag{7.139}$$

is satisfied by infinitely many eigenfunctions, *the characteristic functions* of the duct, each corresponding to a certain discrete, real, nonnegative value of k^2 denoted k_m^2:

$$\left(\frac{\partial^2}{\partial x^2} + \frac{\partial^2}{\partial y^2} \right) \psi_m(x, y) + k_m^2 \psi_m(x, y) = 0. \tag{7.140}$$

The lowest value of k_m^2 is zero, corresponding to the trivial case of ψ_m being independent of x and y. This corresponds to the one-dimensional sound field assumed in Sections 7.2 and 7.3. Combining Equations (7.136) and (7.138) leads to the equation

$$k^2 - k_m^2 + \frac{1}{p_z(z)} \frac{d^2 p_z(z)}{dz^2} = 0, \tag{7.141}$$

which obviously has the solution

$$p_z(z) = A_m e^{-jk_{zm} z} + B_m e^{jk_{zm} z}, \tag{7.142}$$

where

$$k_{zm} = \sqrt{k^2 - k_m^2}. \tag{7.143}$$

Accordingly, the general solution to Equation (7.136) is a sum of waves of the form

$$\hat{p}_m(x, y, z, t) = P_{m+} \psi_m(x, y) e^{j(\omega t - k_{zm} z)} + P_{m-} \psi_m(x, y) e^{j(\omega t + k_{zm} z)}. \tag{7.144}$$

Each mode has a cutoff frequency determined by the eigenvalue k_m^2,

$$f_m = \frac{k_m c}{2\pi}. \tag{7.145}$$

If driven above this frequency the mode can propagate unattenuated; if driven at a lower frequency the mode decays exponentially with a propagation coefficient given by Equation (7.131).[4]

It is apparent that Equation (7.144) is the general version of Equations (7.90) and (7.124). However, in the general case the cross-sectional eigenfunctions ψ_m and the corresponding eigenvalues k_m^2 must be determined numerically, for instance by means of the finite element method.

Example 7.20 An evanescent mode in a duct with arbitrary cross section: A finite element solution of the two-dimensional eigenvalue problem

$$\nabla^2 p(x, y) + k^2 p(x, y) = 0$$

in the 'two-dimensional room' shown in Figure 7.34 has resulted in eigenvalues corresponding to the resonance frequencies 188 Hz, 287 Hz, 330 Hz and 393 Hz. If a duct with this cross-sectional shape is driven at 200 Hz only the plane wave and the mode with a cutoff frequency of 188 Hz can propagate. The phase velocity of the latter is

$$c_{\mathrm{p}} = c/\sqrt{1 - (188/200)^2} \simeq 1005 \, \mathrm{ms}^{-1}.$$

The propagation coefficient of the first non-propagating mode is

$$\gamma_z = k\sqrt{(287/200)^2 - 1} \simeq 3.77 \, \mathrm{m}^{-1}.$$

This mode will have decreased 60 dB after travelling the distance l, where l follows from the relation

$$20 \log e^{\gamma_z l} = 60 \, (\mathrm{dB}),$$

from which we conclude that

$$l = 3 \ln 10/\gamma_z \simeq 1.83 \, \mathrm{m}.$$

7.6 The Green's Function in a Semi-infinite Duct

An interesting general property of the cross-sectional eigenfunctions in a duct can be deduced as follows. From Equation (7.138) we conclude that since

$$\nabla^2 \psi_m + k_m^2 \psi_m = \nabla^2 \psi_n^* + k_n^2 \psi_n^* = 0, \tag{7.146}$$

we also have

$$\psi_n^*(\nabla^2 + k_m^2)\psi_m - \psi_m(\nabla^2 + k_n^2)\psi_n^* = 0. \tag{7.147}$$

---

[4] It can happen that several eigenfunctions have the same eigenvalue k_m^2. This phenomenon is called modal degeneracy. In this case Equation (7.140) will be satisfied by any combination of these eigenfunctions. Spinning modes in ducts with circular cross section are degenerate.

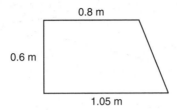

Figure 7.34   Cross section of a uniform duct with rigid walls

However, this equation can be written

$$\nabla \cdot (\psi_n^* \nabla \psi_m - \psi_m \nabla \psi_n^*) + (k_m^2 - k_n^2)\psi_m \psi_n^* = 0. \tag{7.148}$$

Integrating over the cross section and applying Gauss's theorem[5] to the first term gives

$$\int_s (\psi_n^* \nabla \psi_m - \psi_m \nabla \psi_n^*) \cdot \mathbf{n} \, ds + (k_m^2 - k_n^2)\int_S \psi_m \psi_n^* \, dS = 0, \tag{7.149}$$

where s is the perimeter of the cross section and $\mathbf{n}$ is the normal vector. At the rigid walls the normal component of the gradient of any eigenfunction is zero, therefore the left-hand term is zero. It now follows that the eigenfunctions are *orthogonal*, that is, that

$$\int_S \psi_m \psi_n^* \, dS = 0, \tag{7.150}$$

unless $m = n$. It is customary to normalise the eigenfunctions so that

$$\frac{1}{S}\int_S \psi_m \psi_n^* \, dS = \delta_{mn}, \tag{7.151}$$

which implies that

$$\int_S |\psi_m|^2 \, dS = S. \tag{7.152}$$

The orthogonality of the eigenfunctions can be used in determining how a given source excites the various modes. Consider a semi-infinite duct driven by a vibrating surface perpendicular to the duct at $z = 0$, as sketched in Figure 7.35. Let $U(x, y)e^{j\omega t}$ be the vibrational velocity of the surface. The boundary condition at $z = 0$ implies that

$$\hat{u}_z(x, y, 0, t) = U(x, y)e^{j\omega t}, \tag{7.153}$$

that is, from Equation (7.144) (ignoring reflected waves since the tube is semi-infinite),

$$U(x, y) = \sum_{m=0}^{\infty} \frac{p_{m+}}{\rho c} \frac{k_{zm}}{k} \psi_m(x, y). \tag{7.154}$$

---

[5] See Equation (6.7).

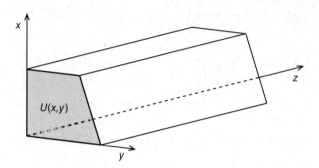

Figure 7.35 Semi-infinite duct excited by a vibrating surface

Multiplication with ψ_n and integration over the cross section give

$$\int_S U(x, y)\psi_n(x, y)\,dS = \sum_{m=0}^{\infty} \frac{p_{m+}}{\rho c}\frac{k_{zm}}{k}\int_S \psi_m(x, y)\psi_n(x, y)\,dS = p_{n+}\frac{S}{\rho c}\frac{k_{zn}}{k}, \quad (7.155)$$

which shows that the amplitude of the m'th term is

$$p_{m+} = \frac{\rho c}{S}\frac{k}{k_{zm}}\int_S U(x, y)\psi_m(x, y)\,dS = \frac{\rho c}{S}\frac{1}{\sqrt{1 - (k_{zm}/k)^2}}\int_S U(x, y)\psi_m(x, y)\,dS.$$
$$(7.156)$$

In the special case where the vibrational velocity is proportional to the shape of a given mode,

$$U(x, y) = U_0\psi_m(x, y), \quad (7.157)$$

only this mode will be generated. (Whether it can propagate or not depends on whether the frequency of the excitation exceeds the cutoff frequency of the mode.) In the general case infinitely many modes will be generated, some propagating and the rest evanescent. Note that the plane wave generated by the vibrating surface has the amplitude

$$p_{0+} = \frac{\rho c}{S}\int_S U(x, y)\,dS = \frac{\rho c}{S}Q, \quad (7.158)$$

where $Qe^{j\omega t}$ is the volume velocity of the vibrating surface (cf. Equation (7.11)); if this surface has no net volume velocity then the plane wave will not be generated.

Example 7.21 Loudspeaker arrangements for driving specific modes: From the foregoing considerations it can be concluded that a loudspeaker mounted in the middle of a plate that terminates a duct with rectangular cross section will generate (m, n)-modes corresponding to even values of m and n. A loudspeaker arrangement consisting of two speakers placed symmetrically as shown in Figure 7.36 and driven in antiphase tends to generate modes corresponding to odd values of m and even values of n.

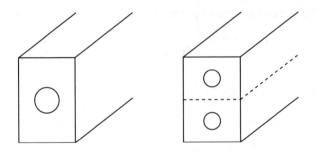

Figure 7.36   Various loudspeaker arrangements for driving a duct

Another important special case occurs if the vibrational pattern is concentrated to a very small region on the surface that terminates the duct. In this case we might regard the source as an infinitely small pulsating sphere of the type described in Section 3.3 but having a finite volume velocity. The source (say, a very small piston), is placed on the surface. Such a source, also known as a point source or a monopole, is studied in detail in Chapter 9. In this case

$$U(x, y) = Q\delta(x - x_0)\delta(y - y_0), \tag{7.159}$$

from which, with Equation (7.156),

$$p_{m+} = \frac{\rho c}{S} \frac{Q}{\sqrt{1 - (k_m/k)^2}} \int_S \psi_m(x, y)\delta(x - x_0)\delta(y - y_0)\,dS = \frac{\rho c}{S} \frac{Q\psi_m(x_0, y_0)}{\sqrt{1 - (k_m/k)^2}}, \tag{7.160}$$

that is, the sound pressure generated by the monopole can be written

$$\hat{p}(x, y, z, t) = \frac{\rho c}{S} Q \sum_{m=0}^{\infty} \frac{\psi_m(x, y)\psi_m(x_0, y_0)}{\sqrt{1 - (k_m/k)^2}} e^{j(\omega t - k_{mz} z)}. \tag{7.161}$$

Note that a given mode is particularly strongly excited if driven just above its cutoff frequency (unless $\psi_m(x_0, y_0)$ is zero, of course). Exactly at the cutoff frequency the sound pressure goes to infinity, but in practice it will, of course, be limited by losses. Note also that the function is symmetrical with respect to source and receiver position; this is a consequence of the general reciprocity relation.

The solution to the inhomogeneous Helmholtz equation

$$(\nabla^2 + k^2)G(\mathbf{r}, \mathbf{r}_0) = -\delta(\mathbf{r} - \mathbf{r}_0) \tag{7.162}$$

(with whatever boundary conditions) is known as the *Green's function*.[6] This is the sound pressure at $\mathbf{r}$ generated by an infinitely small pulsating sphere with a finite volume velocity (a point source or a monopole) at $\mathbf{r}_0$, normalised by letting $j\omega\rho Q = 1$. Apart from the normalisation Equation (7.162) follows if we add a source term (a small pulsating sphere with a finite volume velocity) on the right-hand side of Equation (2.12),

$$\nabla \cdot \mathbf{u}(\mathbf{r}, t) + \frac{1}{\rho c^2} \frac{\partial p(\mathbf{r}, t)}{\partial t} = Q(t)\delta(\mathbf{r} - \mathbf{r}_0), \tag{7.163}$$

---

[6] Some authors have $-4\pi\delta(\mathbf{r} - \mathbf{r}_0)$ on the right-hand side of Equation (7.162).

and combine with the divergence of Equation (2.13),

$$\nabla^2 p(\mathbf{r}, t) + \rho \nabla \cdot \frac{\partial \mathbf{u}(\mathbf{r}, t)}{\partial t} = \nabla^2 p(\mathbf{r}, t) - \frac{1}{c^2} \frac{\partial^2 p(\mathbf{r}, t)}{\partial t^2} + \rho \frac{\partial Q(t)}{\partial t} \delta(\mathbf{r} - \mathbf{r}_0) = 0.$$

(7.164)

For harmonic time dependence using complex representation this expression takes the form

$$\nabla^2 \hat{p}(\mathbf{r}, t) + k^2 \hat{p}(\mathbf{r}, t) = -j\omega\rho Q \delta(\mathbf{r} - \mathbf{r}_0) e^{j\omega t}.$$

(7.165)

We can now conclude that the Green's function is the frequency response between a point source of constant volume *acceleration* at the position $\mathbf{r}_0$ and the resulting sound pressure at the position $\mathbf{r}$, and the Green's function multiplied by $j\omega\rho Q$ is the real physical sound pressure (in pascal). The Green's function is much used in theoretical acoustics because it makes it possible to express the solution to the problem of calculating the sound field generated by a continuity of point sources as an integral. The Green's function for the semi-infinite duct is, from Equation (7.161),

$$G(\mathbf{r}, \mathbf{r}_0) = -\frac{j}{Sk} \sum_{m=0}^{\infty} \frac{\psi_m(x, y) \psi_m(x_0, y_0)}{\sqrt{1 - (k_m/k)^2}}.$$

(7.166)

The magnitude of this function is shown in Figure 7.37 as a function of the frequency. Below the lowest cutoff frequency the sound field is simple because the plane wave dominates. At higher frequencies many modes contribute to the sound field, and since they propagate with different phase velocities complicated interference effects occur. Each higher-order mode is strongly excited at and just above its cutoff frequency.

Figure 7.38 shows the magnitude of the Green's function as a function of the receiver position at a frequency where only the plane wave can propagate and at a higher frequency where many modes contribute. Note the decay of the evanescent mode close to the source.

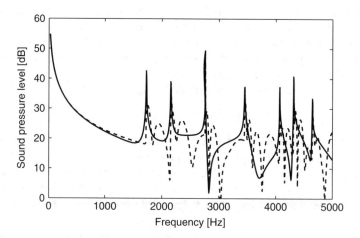

Figure 7.37 Magnitude of the Green's function in a duct as a function of the frequency. Solid line, source and receiver point close to each other; dashed line, source and receiver point far from each other

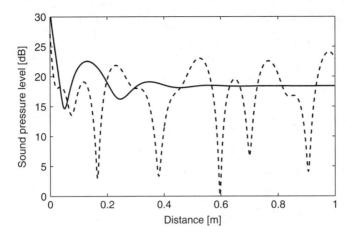

Figure 7.38 Magnitude of the Green's function in a duct as a function of the receiver position. Solid line, just below the lowest cutoff frequency; dashed line, at a frequency where five modes of higher order can propagate

---

Example 7.22 A semi-infinite duct driven by a piston smaller than the cross section: The sound pressure in a duct of rectangular cross section driven by a piston with an area of $ab/2$ as shown in Figure 7.39 can be calculated by integrating the Green's function given by Equation (7.166) over the area of the piston:

$$\hat{p}(x, y, z) = j\omega\rho U \int_{S'} G(\mathbf{r}, \mathbf{r}_0)\,\mathrm{d}S$$

$$= \rho c U \sum_{m=0}^{\infty} \sum_{n=0}^{\infty} \varepsilon_m \varepsilon_n \cos\left(\frac{m\pi x}{a}\right) \cos\left(\frac{n\pi y}{b}\right) \frac{1}{a} \int_0^{a/2} \cos\left(\frac{m\pi x_0}{a}\right) \mathrm{d}x_0$$

$$\times \frac{1}{b} \int_0^b \cos\left(\frac{n\pi y_0}{b}\right) \mathrm{d}y_0\, e^{-jk_{zmn}z}$$

$$= \rho c U \sum_{m=0}^{\infty} \sum_{n=0}^{\infty} \varepsilon_m \varepsilon_n \cos\left(\frac{m\pi x}{a}\right) \cos\left(\frac{n\pi y}{b}\right) \frac{\sin(m\pi/2)}{m\pi} \delta_{n0} e^{-jk_{zmn}z}$$

$$= \rho c U \sum_{m=0}^{\infty} \varepsilon_m \cos\left(\frac{m\pi x}{a}\right) \frac{\sin(m\pi/2)}{m\pi} e^{-jk_{zmn}z}.$$

Note that modes for which $m > 0$ and even and modes for which $n > 0$ are not excited.

We may also make use of the orthogonality of the modes to determine the flow of sound energy in the tube. With

$$\hat{p}(x, y, z, t) = \sum_{m=0}^{\infty} \psi_m(x, y) p_{m+} e^{j(\omega t - k_{zm}z)} \tag{7.167}$$

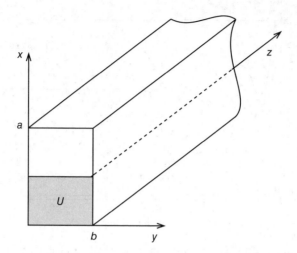

Figure 7.39 A semi-infinite duct driven by a piston with an area of $ab/2$

the particle velocity in the axial direction is

$$\hat{u}_z(x, y, z, t) = \sum_{m=0}^{\infty} \psi_m(x, y) \frac{k_{zm}}{k} \frac{p_{m+}}{\rho c} e^{j(\omega t - k_{zm} z)}. \tag{7.168}$$

It now follows that the sound power transmitted through the duct is

$$
\begin{aligned}
P_a &= \int_S I_z \, \mathrm{d}S = \int_S \frac{1}{2} \mathrm{Re}\{\hat{p} \hat{u}_z^*\} \, \mathrm{d}S \\
&= \frac{1}{2\rho c} \int_S \mathrm{Re} \left\{ \left(\sum_{m=0}^{\infty} \psi_m(x, y) \, p_{m+} e^{-jk_{zm} z} \right) \left(\sum_{n=0}^{\infty} \psi_n^*(x, y) \frac{k_{zn}^*}{k} p_{n+}^* e^{jk_{zn} z} \right) \right\} \, \mathrm{d}S.
\end{aligned}
\tag{7.169}
$$

However, because of the orthogonality of the eigenfunctions all cross terms $(m = n)$ integrate to zero, which means that Equation (7.169) can be reduced to

$$P_a = \frac{1}{2\rho c} \int_S \left(\sum_{m=0}^{\infty} |\psi_m(x, y)|^2 \frac{\mathrm{Re}\{k_{zm}\}}{k} |p_{m+}|^2 \right) \mathrm{d}S = \frac{S}{2\rho c k} \sum_{m=0}^{\infty} |p_{m+}|^2 \mathrm{Re}\{k_{zm}\}, \tag{7.170}$$

which shows that the transmitted power is simply the sum of the powers of the individual modes. Only the propagating modes (for which k_{zm} is real-valued) contribute to this sum.

7.7 Sound Propagation in Ducts with Walls of Finite Impedance

If the walls of the duct can be assumed to be locally reacting with the specific acoustic impedance $Z_{s,w}$ (see Section 5.3) the boundary condition at the walls is

$$\frac{\hat{p}}{\hat{u}_n} = Z_{s,w}, \tag{7.171}$$

where u_n is the normal component of the particle velocity. For simplicity we will restrict the analysis to axisymmetric modes in ducts of circular cross section. From Equation (7.124) with $m = 0$ it follows that the sound pressure can be written in the form

$$\hat{p}(r, z, t) = AJ_0(k_r r)e^{j(\omega t - k_z z)}. \tag{7.172}$$

The particle velocity in the radial direction is

$$\hat{u}_r(r, z, t) = -\frac{1}{j\omega\rho}\frac{\partial \hat{p}}{\partial r} = -\frac{Ak_r}{j\omega\rho}J_0'(k_r r)e^{j(\omega t - k_z z)}$$
$$= \frac{Ak_r}{j\rho ck}J_1(k_r r)e^{j(\omega t - k_z z)}, \tag{7.173}$$

where we have used the fact that $J_0'(z) = -J_1(z)$. Combining Equations (7.171), (7.172) and (7.173) gives the *eigen equation*

$$k_r J_1(k_r a) = jk\frac{\rho c}{Z_{s,w}}J_0(k_r a). \tag{7.174}$$

This equation is satisfied for certain discrete values of k_r, denoted k_{rn} with $n = 0$, 1, 2, ... With an infinitely large wall impedance corresponding to a duct with perfectly rigid walls, $k_{rn}a$ are the roots of $J_1(z)$ (0, 3.832, 7.016, ...) found in Section 7.5.2. However, in the general case the transcendental equation (7.174) must be solved numerically for k_r, which in general is complex unless $Z_{s,w}$ is purely imaginary. The axial wavenumber can be calculated from Equation (7.143),

$$k_{zn} = \sqrt{k^2 - k_{rn}^2}, \tag{7.175}$$

and this will also be complex unless $Z_{s,w}$ is purely imaginary, indicating that all modes decay exponentially. This agrees with the fact that the real part of the wall impedance is associated with losses.

7.7.1 Ducts with Nearly Hard Walls

If $|Z_{s,w}| \gg \rho c$ the discrete values of k_r that satisfy Equation (7.174) must be close to the values of k_r for ducts with rigid walls; therefore they can be found at least approximately from truncated series of the Bessel functions.

For small arguments, $J_1(z) = z/2$ and $J_0(0) = 1$ (see Appendix C), and Equation (7.174) becomes

$$\frac{k_{r0}^2}{2}a \simeq jk\frac{\rho c}{Z_{s,w}}, \tag{7.176}$$

from which it follows that the radial wavenumber of the 'plane' wave is

$$k_{r0} \simeq \left(\frac{j2k\rho c}{aZ_{s,w}}\right)^{1/2}. \tag{7.177}$$

The corresponding axial wavenumber is

$$k_{z0} = \sqrt{k^2 - k_{r0}^2} = k\left(1 - \frac{2j\rho c}{ka Z_{s,w}}\right)^{1/2} \simeq k\left(1 - \frac{j\rho c}{ka Z_{s,w}}\right). \tag{7.178}$$

It can now be seen that the sound pressure varies slightly with the radius even in the fundamental mode, which strictly is no longer a plane wave. It can also be concluded that the wave decays at the rate

$$\frac{20}{\ln 10} \mathrm{Re}\left\{\frac{\rho c}{Z_{s,w}}\right\}\frac{1}{a} \simeq 8.69 \mathrm{Re}\{\beta\}/a \qquad \text{(dB per unit length).} \qquad (7.179)$$

where

$$\beta = \rho c / Z_{s,w} \qquad (7.180)$$

is the normalised admittance of the wall. Another interesting conclusion to be drawn from Equation (7.178) is that the phase speed is lower than the free field speed of sound in a tube with compliant walls, that is, walls for which the imaginary part of the normalised admittance is positive. From Equations (7.129) and (7.178) it can be seen that

$$c_p = \frac{\omega}{\mathrm{Re}\{k_{z0}\}} = \frac{\omega}{k + \mathrm{Im}\{\beta\}/a} = \frac{c}{1 + \mathrm{Im}\{\beta\}/ka}. \qquad (7.181)$$

Conversely, the phase speed exceeds the speed of sound if the walls are mass-like.

Modes of higher order are affected in a similar manner. The perturbed values of k_{rn} and k_{zn} can be determined by expanding the Bessel functions of the eigen equation into Taylor series at the roots of $J_1(z)$. One consequence of a finite real part of the wall admittance is that there is no sharp cutoff frequency associated with each higher-order mode, since all modes decay to some extent at all frequencies.

Even in a duct with perfectly rigid walls there are some losses, which we have ignored up to now. In a thin boundary layer very near the walls heat conduction and viscosity cannot be ignored. The effect of these phenomena can be described in terms of a small but finite wall admittance,

$$\beta = \frac{1+j}{2c}(\sqrt{2v\omega} + (\gamma - 1)\sqrt{2\alpha_t\omega}), \qquad (7.182)$$

where γ is the ratio of the specific heats (cf. Equation (2.8)), v is the kinematic viscosity ($\simeq 15.2 \cdot 10^{-6}\,\mathrm{m}^2/\mathrm{s}$ for air), and α_t is the thermal diffusivity ($\simeq 21.2 \cdot 10^{-6}\,\mathrm{m}^2/\mathrm{s}$ for air) [9]. From Equation (7.182) we conclude that a plane wave in a tube with rigid walls decays at the rate

$$\frac{20}{\ln 10}\mathrm{Re}\{\beta\}/a \simeq 8.69\frac{\sqrt{v\omega}}{ac}\frac{1 + (\gamma - 1)\sqrt{\alpha_t/v}}{\sqrt{2}} \qquad \text{(dB per unit length),} \qquad (7.183)$$

because of thermal losses and viscous effects. Note that the rate of decay is proportional to the square root of the frequency and inversely proportional to the radius of the duct.

Example 7.23 Attenuation of sound due to viscous and thermal losses in a duct: At 1 kHz a sound wave in a duct with radius 1 cm decays 0.8 dB per metre.

7.7.2   Lined Ducts

Unless axial sound propagation in the lining material is prevented by solid partitions the assumption that the lining is of local reaction is not very realistic. When axial propagation is prevented the attenuation of sound waves in the lining will be particularly strong when the thickness of the lining equals an odd-numbered multiple of a quarter of a wavelength (in the lining material). However, partitions to prevent axial propagation are rarely used. A more realistic assumption in most cases is that the lining is *bulk reacting*, which implies that sound propagation in the lining can be described in terms of a complex wavenumber and a complex characteristic impedance. Simple empirical expressions for these quantities expressed in terms of the flow resistance of the material are available in the literature [10]. Continuity of the sound pressure and of the normal component of the particle velocity at the interface leads to the assumption of a common propagation coefficient. The result of the analysis is an equation similar to (although more complicated than) Equation (7.176). This equation must be solved numerically.

References

[1] A.D. Pierce: *Acoustics. An Introduction to Its Physical Principles and Applications*. The American Institute of Physics, New York (1989). See Section 5.4.

[2] H. Levine and J. Schwinger: On the radiation of sound from an unflanged circular pipe. *Physical Review* **73**, 383–406 (1948).

[3] T.D. Rossing, R.F. Moore and P.A. Wheeler: *The Science of Sound* (3rd edition). Addison Wesley, San Francisco, CA (2002).

[4] P. Choudary Chaitanaya and M.L. Munjal: Tuning of the extended concentric tube resonators. *International Journal of Acoustics and Vibration* **16**, 111–118 (2011).

[5] D.A. Bies and C.H. Hansen: *Engineering Noise Control* (2nd edition). E & FN Spon, London, 1996.

[6] C.L. Morfey: Acoustic energy in non-uniform flow. *Journal of Sound and Vibration* **14**, 159–170 (1971).

[7] A. Papoulis: *Signal Analysis*. McGraw-Hill, New York, 1977. See Section 4.2.

[8] E.G. Williams: *Fourier Acoustics. Sound Radiation and Nearfield Acoustic Holography*. Academic Press, San Diego, 1999.

[9] C. Zwicker and C.W. Kosten: *Sound Absorbing Materials*. Elsevier Publishing Company, New York (1949).

[10] M.E. Delany and E.N. Bazley: Acoustical properties of fibrous absorbent materials. *Applied Acoustics* **3**, 105–116 (1970).

8

Sound in Enclosures

8.1 Introduction

Sound in enclosures is a wide field since 'enclosures' (volumes enclosed by solid surfaces) range from small acoustic couplers for calibrating microphones to large rooms such as auditoria and concert halls where perception of sound is essential.

The study of sound fields in undamped or lightly damped enclosed spaces can be approached in at least two completely different ways. One can solve the wave equation with the prescribed boundary conditions either analytically or using numerical methods. This approach, which is particularly useful in enclosures with characteristic dimensions comparable to or shorter than the wavelength, leads to a description in terms of the modes of the enclosure. Alternatively the problem can be studied using statistical considerations. The second approach, which is particularly appropriate in rooms at medium and high frequencies, has the advantage of requiring far less detailed knowledge of the geometry of the room under study, but the resulting model is not very accurate at low frequencies.

The purpose of this chapter is to give an elementary introduction to these two models. Other important topics in room acoustics such as ray acoustic models, image source models, models based on the diffusion theory, and subjective parameters of importance for perception of sound in rooms are *not* dealt with; the reader is referred to the list of books in the bibliography.

8.2 The Modal Theory of Sound in Enclosures

Sound in lightly damped enclosures will resonate at certain frequencies, the 'natural frequencies', and they are usually well separated at low frequencies. The sound field in the enclosure at such a resonance frequency is called a 'mode' (or, for reasons that will become apparent, a 'normal mode'), and the spatial distribution of the sound pressure is called the 'mode shape'. If losses at the walls are ignored the Helmholtz equation with the boundary conditions imposed by the rigid walls becomes an eigenvalue problem, and the mathematical solution of the equation leads to eigenfunctions and eigenfrequencies, which are mathematical terms for the modes and the natural frequencies.

Fundamentals of General Linear Acoustics, First Edition. Finn Jacobsen and Peter Møller Juhl.
© 2013 John Wiley & Sons, Ltd. Published 2013 by John Wiley & Sons, Ltd.

The modes and natural frequencies cannot be determined analytically unless the enclosure is of a simple shape. In enclosures of a more complicated geometry the solution must be found using numerical methods, such as for example the finite element method.

8.2.1 Eigenfrequencies and Mode Shapes

For simplicity we will concentrate on the particularly simple case of a rectangular enclosure. This means that our task is to find solutions to the Helmholtz equation expressed in the usual Cartesian coordinate system,

$$\frac{\partial^2 \hat{p}}{\partial x^2} + \frac{\partial^2 \hat{p}}{\partial y^2} + \frac{\partial^2 \hat{p}}{\partial z^2} + k^2 \hat{p} = 0, \tag{8.1}$$

subject to the boundary conditions of zero normal sound pressure gradient at the rigid walls,

$$\frac{\partial \hat{p}}{\partial x} = 0 \text{ at } x = \begin{cases} 0 \\ l_x \end{cases} ; \quad \frac{\partial \hat{p}}{\partial y} = 0 \text{ at } y = \begin{cases} 0 \\ l_y \end{cases} ; \quad \frac{\partial \hat{p}}{\partial z} = 0 \text{ at } z = \begin{cases} 0 \\ l_z \end{cases} . \tag{8.2a, 8.2b, 8.2c}$$

The problem is obviously a special case of the problem studied in Section 7.5, where it was shown that the general solution can be written in the form

$$\hat{p} = (A e^{-jk_x x} + B e^{jk_x x})(C e^{-jk_y y} + D e^{jk_y y})(E e^{-jk_z z} + F e^{jk_z z}) e^{j\omega t}, \tag{8.3}$$

where

$$k_x^2 + k_y^2 + k_z^2 = k^2. \tag{8.4}$$

From the boundary conditions we conclude that $A = B$, $C = D$ and $E = F$, and that

$$k_x l_x = n_x \pi, \quad k_y l_y = n_y \pi, \quad k_z l_z = n_z \pi, \tag{8.5a, 8.5b, 8.5c}$$

where n_x, n_y and n_z are integers (cf. Equations (7.87) and (7.88)). All in all we can write the sound pressure as a sum of terms of the form

$$\psi_N(x, y, z) = \sqrt{\varepsilon_{n_x} \varepsilon_{n_y} \varepsilon_{n_z}} \cos\left(\frac{n_x \pi x}{l_x}\right) \cos\left(\frac{n_y \pi y}{l_y}\right) \cos\left(\frac{n_z \pi z}{l_z}\right), \tag{8.6}$$

that is,

$$\hat{p}(x, y, z, t) = \sum_N A_N \psi_N(x, y, z) e^{j\omega t}, \tag{8.7}$$

where N represents the three integers n_x, n_y and n_z,

$$\sum_N = \sum_{n_x=0}^{\infty} \sum_{n_y=0}^{\infty} \sum_{n_z=0}^{\infty}. \tag{8.8}$$

As shown in Section 8.2.3 the modal amplitudes, A_N, depend on the position and strength of the source; and the factor $\sqrt{\varepsilon_{n_x} \varepsilon_{n_y} \varepsilon_{n_z}}$ is a normalisation constant (cf. Equation (7.90)).

Combining Equations (8.4) and (8.5) gives us an expression for the natural frequencies of the room (in hertz),[1]

$$f_N = \frac{\omega_N}{2\pi} = \frac{k_N c}{2\pi} = \frac{c}{2}\left(\left(\frac{n_x}{l_x}\right)^2 + \left(\frac{n_y}{l_y}\right)^2 + \left(\frac{n_z}{l_z}\right)^2\right)^{1/2}. \tag{8.9}$$

Each term in the sum given by Equation (8.7) represents a mode. In the special case where all three indices are zero we have the fundamental 'cavity mode' in which the sound pressure is independent of the position in the room and the air in the room acts like a spring (see Example 8.3). If two out of the three indices are zero, we have an *axial mode* with wave motion in just one direction; two-dimensional modes, for which one index equals zero, are known as *tangential modes*; and three-dimensional modes are also called *oblique modes*. In Figure 8.1 are shown contours of equal sound pressure in two different tangential modes. Note the nodal planes in which the sound pressure is zero.

Modes in a room may be interpreted as sums of interfering waves travelling in various directions. For example, if we rewrite the expression for an axial mode in the x-direction as a sum of two complex exponentials,

$$\psi_N(x, y, z) = \frac{1}{\sqrt{2}}(e^{-jk_x x} + e^{jk_x x}), \tag{8.10}$$

it becomes apparent that it may be regarded as a sum of two plane waves with the same amplitude, one travelling in the positive x-direction and one travelling in the opposite direction – in other words a standing wave. The two waves interfere in such a manner that the boundary conditions at $x = 0$ and $x = l_x$ are satisfied; this is the case only for the discrete values of k_x given by Equation (8.5a). In the same way we can write the

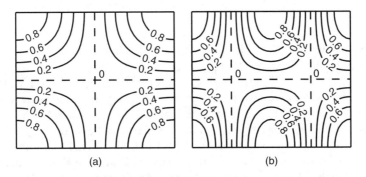

(a)   (b)

Figure 8.1 Equal pressure contours of the (1, 1, 0) mode and the (2, 1, 0) mode in a rectangular room

---

[1] If the dimensions of the room (or some of them) are commensurable (say, $l_x = 2l_y$) some of the natural frequencies will coincide. As a result the frequency response of the room will be more irregular. This phenomenon, which is called modal degeneracy, should be avoided. The dimensions of well-designed rectangular reverberation rooms (or loudspeaker enclosures) are not related by whole numbers.

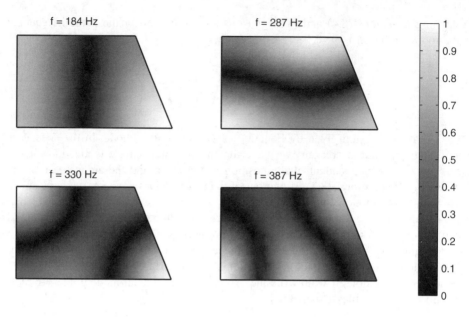

Figure 8.2   Modal patterns in a nonrectangular room

general expression for a three-dimensional mode in the form

$$\Psi_N(x, y, z) = \frac{1}{\sqrt{8}} (e^{-jk_x x} + e^{jk_x x})(e^{-jk_y y} + e^{jk_y y})(e^{-jk_z z} + e^{jk_z z})$$

$$= \frac{1}{\sqrt{8}} (e^{-j(k_x x + k_y y + k_z z)} + e^{-j(k_x x + k_y y - k_z z)} + e^{-j(k_x x - k_y y + k_z z)} + e^{-j(k_x x - k_y y - k_z z)}$$

$$+ e^{j(k_x x + k_y y + k_z z)} + e^{j(k_x x + k_y y - k_z z)} + e^{j(k_x x - k_y y + k_z z)} + e^{j(k_x x - k_y y - k_z z)}),$$

$$(8.11)$$

which shows that an oblique mode can be decomposed into eight interfering plane waves that propagate in the directions (k_x, k_y, k_z), $(k_x, k_y, -k_z)$, $(k_x, -k_y, k_z)$, $(k_x, -k_y, -k_z)$, $(-k_x, -k_y, -k_z)$, $(-k_x, -k_y, k_z)$, $(-k_x, k_y, -k_z)$ and $(-k_x, k_y, k_z)$ (cf. Equation (7.97)). These plane waves may be interpreted as one obliquely incident plane wave and its reflections from the walls of the room. The eight waves interfere in such a manner that the boundary conditions (Equations (8.2a, 8.2b, 8.2c)) are satisfied.

It should finally be emphasised that the sound field in an enclosure of *any* shape can be decomposed into modes, although simple analytical solutions are available only for rectangular, cylindrical and spherical enclosures. An example of a mode in an irregular room is given in Figure 8.2. In such a room there are no nodal planes but curved nodal surfaces.

Example 8.1 Modes in a cylindrical cavity with circular cross section: The modes in a cylindrical enclosure with rigid walls at $z = 0$, $z = h$ and $r = a$ have the general form

$$\psi_{lmn}(r, \varphi, z) = \Lambda_{mn}\sqrt{\varepsilon_l \varepsilon_m} J_m(k_{rmn}r)\cos(m\varphi + \varphi_m)\cos(k_{zl}z),$$

where the radial wavenumber k_{rmn} is determined by Equation (7.123), and the axial wavenumber is

$$k_{zl} = \frac{l\pi}{h}$$

with $l = 1, 2, 3, \ldots$. The eigenfrequencies follow:

$$f_{mnl} = \frac{k_{lmn}c}{2\pi} = \frac{c}{2}\sqrt{\left(\frac{l}{h}\right)^2 + \left(\frac{\eta_{mn}}{a\pi}\right)^2},$$

where η_{mn} are roots of Equation (7.123). The modes can be classified as longitudinal modes ($\eta_{mn} = 0; l > 0$), axisymmetric ($m = 0$) or non-axisymmetric ($m > 0$) transversal modes ($\eta_{mn} > 0, l = 0$), and three-dimensional modes ($\eta_{mn} > 0, l > 0$).

Example 8.2 Modes in a cylindrical cavity of arbitrary cross section: A duct of arbitrary cross section is terminated by rigid walls at $z = 0$ and at $z = h$. Equations (7.138) and (8.5c) lead to the conclusion that the modes of the resulting cavity can be written

$$\psi_{mn}(x, y, z) = O_m\sqrt{\varepsilon_n}\psi_m(x, y)\cos(k_n z),$$

where ψ_m is a solution to the two-dimensional eigenvalue problem given by Equation (7.138) and

$$k_n = \frac{n\pi}{h}$$

(cf. Equation (8.5) and Example 8.1). The quantity ε_n is the usual normalisation factor (cf. Equation (7.90)), and O_m is another normalisation factor that depends on ψ_m. The eigenfrequencies of the cavity are given by

$$f_{mn} = \frac{k_{mn}c}{2\pi} = \frac{c}{2}\sqrt{\left(\frac{k_m}{\pi}\right)^2 + \left(\frac{n}{h}\right)^2},$$

where k_m is the eigenvalue of the two-dimensional problem defined by the cross section.

8.2.2 The Modal Density

The exact distribution of the eigenfrequencies of a rectangular room depends on the dimensions as indicated by Equation (8.9). The modal density, that is, the number of modes per unit bandwidth, is in general a somewhat irregular function of the frequency. However, it is possible to derive a smoothed expression for the modal density, as follows. From the observation that the natural frequencies of the axial modes in the x-direction are equidistantly distributed on the frequency axis (in hertz),

$$\frac{c}{2l_x}, \quad \frac{c}{l_x}, \quad \frac{3c}{2l_x}, \quad \frac{2c}{l_x}, \quad \cdots \tag{8.12}$$

we conclude that there are about $f/(c/2l_x)$ such modes below the frequency f. In a similar manner we can derive an approximate expression for the number of tangential $x - y$ modes below the frequency f by counting the number of rectangles with dimensions $(c/2l_x, c/2l_y)$ in a quarter of a circle with radius f (see Figure 8.3). The result is

$$N_{xy}(f) \simeq \frac{\frac{1}{4}\pi f^2}{\frac{c}{2l_x}\frac{c}{2l_y}} = \frac{\pi l_x l_y}{c^2} f^2. \tag{8.13}$$

Finally we can derive an approximate expression for the number of three-dimensional modes below f by counting the number of rectangular boxes with dimensions $(c/2l_x, c/2l_y, c/2l_z)$ in an eighth of a sphere with the radius f,

$$N(f) \simeq \frac{\frac{1}{8}\frac{4}{3}\pi f^3}{\frac{c}{2l_x}\frac{c}{2l_y}\frac{c}{2l_z}} = \frac{4\pi V}{3c^3} f^3. \tag{8.14}$$

The observation that the number of one-, two-, and three-dimensional modes below f is proportional to f, f^2 and f^3 leads to the conclusion that three-dimensional modes will dominate except at low frequencies; therefore we might ignore axial and tangential modes. Differentiating Equation (8.14) with respect to frequency gives the modal density (in modes per hertz),

$$n(f) = \frac{\mathrm{d}N(f)}{\mathrm{d}f} \simeq \frac{4\pi V}{c^3} f^2. \tag{8.15}$$

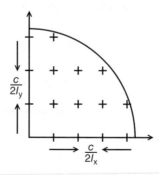

Figure 8.3   Counting the number of $x - y$ modes below the frequency f

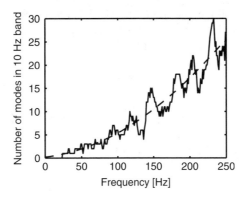

Figure 8.4 Number of natural frequencies in a 10-Hz band compared with the smoothed modal density

It can be shown that this expression is asymptotically valid in any room, irrespective of its shape [1]. A more careful derivation for the special case of rectangular enclosures involving small corrections to Equations (8.13) and (8.14) leads to an expression of the form

$$n(f) = \frac{4\pi V}{c^3} f^2 + \frac{\pi S}{2c^3} f + \frac{L}{8c}, \tag{8.16}$$

where $S = 2(l_x l_y + l_x l_z + l_y l_z)$ is the surface area of the room and $L = 4(l_x + l_y + l_z)$ [2]. This expression multiplied by a bandwidth of 10 Hz is compared with the actual number of modes per 10 Hz (an integer) in Figure 8.4.

8.2.3 The Green's Function in an Enclosure

An important property of the modes in a room can be deduced as follows. Since every mode satisfies the Helmholtz equation, we can write

$$\nabla^2 \psi_m + k_m^2 \psi_m = 0; \; \nabla^2 \psi_n + k_n^2 \psi_n = 0, \tag{8.17a, 8.17b}$$

and therefore,

$$\psi_n (\nabla^2 + k_m^2) \psi_m - \psi_m (\nabla^2 + k_n^2) \psi_n = 0. \tag{8.18}$$

However, this equation can be rewritten in the form

$$\nabla \cdot (\psi_n \nabla \psi_m - \psi_m \nabla \psi_n) + (k_m^2 - k_n^2) \psi_m \psi_n = 0. \tag{8.19}$$

Integrating over the volume of the room and applying Gauss's theorem[2] on the first term gives

$$\int_S \left(\psi_n \frac{\partial \psi_m}{\partial n} - \psi_m \frac{\partial \psi_n}{\partial n} \right) dS + (k_m^2 - k_n^2) \int_V \psi_m \psi_n dV = 0, \tag{8.20}$$

---

[2] See Equation (6.7).

where S is the surface of the room. At the rigid walls the normal component of the gradient of any eigenfunction is zero, and therefore the left-hand term is zero. It follows that the eigenfunctions are *orthogonal*, that is, that

$$\int_V \psi_m \psi_n dV = 0, \tag{8.21}$$

unless $m = n$; hence the term 'normal mode'. It is customary to normalise the eigenfunctions so that

$$\frac{1}{V} \int_V \psi_m \psi_n dV = \delta_{mn}, \tag{8.22}$$

which implies that

$$\int_V \psi_m^2 dV = V. \tag{8.23}$$

Note that these considerations have *not* been limited to the special case of a rectangular room.

It is interesting to study how a source placed at a certain position in the room will excite the various modes. The Green's function for the sound field in a room with rigid walls can be derived as follows. We are looking for solutions to the inhomogeneous Helmholtz equation

$$\nabla^2 G(\mathbf{r}, \mathbf{r_0}) + k^2 G(\mathbf{r}, \mathbf{r_0}) = -\delta(\mathbf{r} - \mathbf{r_0}) \tag{8.24}$$

with the boundary condition

$$\frac{\partial G(\mathbf{r}, \mathbf{r_0})}{\partial n} = 0 \tag{8.25}$$

on the walls (cf. Equation (7.162)). Any sound field in the room can be expressed in terms of the modes of the room, that is, functions that satisfy the equation

$$\nabla^2 \psi_m(\mathbf{r}) + k_m^2 \psi_m(\mathbf{r}) = 0 \tag{8.26}$$

and the boundary condition mentioned above. Therefore,

$$G(\mathbf{r}, \mathbf{r_0}) = \sum_m A_m \psi_m(\mathbf{r}). \tag{8.27}$$

The source term (the right-hand side of Equation (8.24)) can also be expanded into a sum of modes,

$$-\delta(\mathbf{r} - \mathbf{r_0}) = \sum_m B_m \psi_m(\mathbf{r}). \tag{8.28}$$

Multiplying with ψ_n and integrating over the volume of the room gives, if we make use of the fact that the modes are orthogonal (cf. Equation (8.22)),

$$-\int_V \delta(\mathbf{r} - \mathbf{r_0}) \psi_n(\mathbf{r}) dV = -\psi_n(\mathbf{r_0}) = \int_V \sum_m B_m \psi_m(\mathbf{r}) \psi_n(\mathbf{r}) dV = B_n V, \tag{8.29}$$

which shows that

$$B_m = -\frac{\psi_m(\mathbf{r_0})}{V}, \tag{8.30}$$

and thus

$$-\delta(\mathbf{r} - \mathbf{r_0}) = -\frac{1}{V}\sum_m \psi_m(\mathbf{r})\psi_m(\mathbf{r_0}).$$

(8.31)

It now follows that

$$(\nabla^2 + k^2)G(\mathbf{r}, \mathbf{r_0}) = (\nabla^2 + k^2)\sum_m A_m \psi_m(\mathbf{r}) = \sum_m A_m(\nabla^2 + k_m^2 - k_m^2 + k^2)\psi_m(\mathbf{r})$$

$$= \sum_m A_m(k^2 - k_m^2)\psi_m(\mathbf{r}) = -\frac{1}{V}\sum_m \psi_m(\mathbf{r})\psi_m(\mathbf{r_0}),$$

(8.32)

from which we deduce that

$$A_m = -\frac{1}{V}\frac{\psi_m(\mathbf{r_0})}{k^2 - k_m^2}.$$

(8.33)

Finally we can write the Green's function as

$$G(\mathbf{r}, \mathbf{r_0}) = -\frac{1}{V}\sum_m \frac{\psi_m(\mathbf{r})\psi_m(\mathbf{r_0})}{k^2 - k_m^2}.$$

(8.34)

Note the symmetry with respect to source and receiver position, in agreement with the reciprocity principle. A point source placed on a nodal surface of a given mode does not excite the mode. Note also that each mode, not surprisingly, contributes most to the sound field when driven near its natural frequency ($k \simeq k_m$). It can be seen that the response is unlimited if the frequency of the excitation coincides with one of the natural frequencies.

In practice there are, of course, losses in any enclosure, even in a room with walls of solid concrete.[3] As a result the eigenvalues of the problem become complex, with small imaginary parts equal to $1/(2\tau_m c)$, where τ_m is the time constant of the m'th mode (see Section 8.4.1). This leads to the expression

$$G(\mathbf{r}, \mathbf{r_0}) \simeq -\frac{1}{V}\sum_m \frac{\psi_m(\mathbf{r})\psi_m(\mathbf{r_0})}{k^2 - k_m^2 - jk_m/(\tau_m c)} \simeq -\frac{1}{V}\sum_m \frac{\psi_m(\mathbf{r})\psi_m(\mathbf{r_0})}{k^2 - k_m^2 - jk/(\tau_m c)}$$

$$= -\frac{c^2}{V}\sum_m \frac{\psi_m(\mathbf{r})\psi_m(\mathbf{r_0})}{\omega^2 - \omega_m^2 - j\omega/\tau_m} = -\frac{c^2}{V(2\pi)^2}\sum_m \frac{\psi_m(\mathbf{r})\psi_m(\mathbf{r_0})}{f^2 - f_m^2 - jf/(2\pi\tau_m)}$$

(8.35)

where the second approximation has the advantage over the first one that it corresponds to a real-valued, causal time function.[4] The first expression corresponds to complex

---

[3] Even in a room with perfectly rigid walls there are some losses because of the viscosity of air and because the process of sound tends to be isothermal rather than adiabatic in a thin boundary layer near every wall. To a rough approximation the effect of the thermal and viscous losses at the walls in a reverberation room can be described in terms of a finite absorption coefficient,

$$\alpha \simeq 1.8 \cdot 10^{-4}\sqrt{f},$$

where f is the frequency in hertz [3]. See Section 8.4 for relations between the absorption coefficient, the wall admittance, the time constant, the reverberation time, etc.

[4] A realistic frequency response should correspond to a real-valued, causal impulse response. This implies that the real part of the frequency response must be an even function of the frequency, the imaginary part should be an odd function of the frequency, and that the real and imaginary part are related by the Hilbert transform (see [4] and Appendix B).

eigenvalues caused by losses at the boundaries; the other versions correspond to a *medium* with losses.

It can be seen from Equation (8.35) that the response of any mode is limited also when it is driven at its natural frequency. It is also easy to show that the 3-dB bandwidth of the m'th mode is $1/\tau_m$ in radians per second and $1/(2\pi\tau_m)$ in hertz. We will study the effect of losses in Section 8.4.

Figure 8.5 shows two examples of Equation (8.35) plotted as a function of the frequency, and Figure 8.6 shows two examples of Equation (8.35) plotted as a function of the distance between the source and the receiver position.

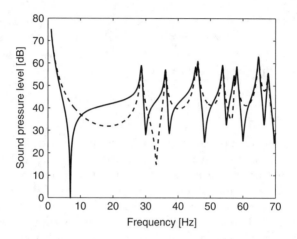

Figure 8.5 Magnitude of the Green's function in a room as a function of the frequency. Solid line: source and receiver points close to each other; dashed line: source and receiver points far from each other

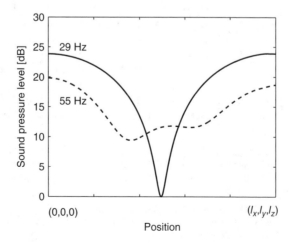

Figure 8.6 Magnitude of the Green's function in a room as a function of the receiver position. The room is driven close to a natural frequency (solid line) and midway between two natural frequencies (dashed line)

Example 8.3 The Green's function in an enclosure at very low frequencies: If the room is driven at a very low frequency the response will be dominated by the $(0, 0, 0)$ mode. Thus Equation (8.35) becomes,

$$G(\mathbf{r}, \mathbf{r_0}) \simeq -\frac{1}{Vk^2} = -\frac{c^2}{V\omega^2},$$

which is independent of $\mathbf{r}$ and $\mathbf{r_0}$. The sound pressure generated by a monopole with the volume velocity $Qe^{j\omega t}$ is obtained by multiplying with $j\omega\rho Q$ (cf. Equations (7.161) and (7.166)),

$$\hat{p}(\mathbf{r}) = j\omega\rho Q G(\mathbf{r}, \mathbf{r_0})e^{j\omega t} \simeq \frac{\rho c^2}{j\omega V}Qe^{j\omega t},$$

and this is seen to agree with the fact that the acoustic impedance of a cavity with dimensions much shorter than the wavelength is

$$Z_a = \frac{\rho c^2}{j\omega V} = \frac{\gamma p_0}{j\omega V}$$

(cf. Equations (5.6) and (5.8)).

The $(0, 0, 0)$ 'cavity mode' is responsible for the low frequency boost that can be heard in small spaces, e.g., inside cars. The acoustic impedance of the cavity is much larger than the radiation impedance in the open, and therefore even a small loudspeaker can generate a surprisingly high sound pressure level above its resonance frequency and up to, say, 80 Hz in a car.

Example 8.4 Comparison with the Green's function in a duct: It is interesting to compare the Green's function in a lossless room (Equation (8.34)) with the earlier derived Green's functions in a duct. The Green's function for a semi-infinite lossless duct is given by Equation (7.166),

$$G(\mathbf{r}, \mathbf{r_0}) = -\frac{j}{S}\sum_m \frac{\psi_m(\mathbf{r})\psi_m(\mathbf{r_0})}{\sqrt{k^2 - k_m^2}}e^{-jk_{zm}z}.$$

There are evidently strong similarities between modes in ducts and rooms, but the cut-off phenomenon that occurs when a duct mode is driven below a certain frequency is peculiar to ducts.

8.3 Statistical Room Acoustics

In theory the validity of the modal expressions derived in Section 8.2 is not restricted to low frequencies. It may seem surprising, therefore, that a completely different approach based on statistical considerations is actually more useful at medium and high frequencies than the deterministic approach described in the foregoing. Evidently, a model based on statistical considerations can only give statistical answers; we may, for example, be able to predict with a certain level of confidence that the sound pressure level is within a certain range. However, we can never make deterministic predictions with a probabilistic model. By contrast, Equation (8.34) appears to be exact. Why the concern with statistical models, then?

There are two reasons. One reason is that expressions based on sums of modes in practice are less useful at high frequencies than they would seem to be. The problem is that when hundreds of complex terms are summed the result becomes very sensitive to small errors in each term, and such errors are likely to occur. For example, the dimensions of the room might be slightly different from the dimensions used in the model, and the temperature and thus the speed of sound may differ a bit from the value that is assumed. Even very small modelling errors will shift the natural frequencies of the modes, and the amplitude and phase of each of the terms that correspond to modes driven near their natural frequency may change somewhat. As a result the sum can be completely wrong at any given frequency.

The other reason is that statistical models, as we shall see, can be surprisingly powerful in the sense that they make it possible to predict a number of characteristics of, e.g., the sound field in a reverberation room on the basis of very little information. Indeed, the statistical properties of the sound field in a reverberation room driven with a pure tone can be predicted in considerable detail without *any* knowledge of the room.

Only some of the most fundamental results of the statistical model will be presented here. The reader is referred to [5, 6] for further details.

The background for the statistical theory is the fact that the modal density in a room to a good approximation increases with the square of the frequency, irrespective of the shape of the room, as we have seen in Section 8.2,

$$n(f) \simeq \frac{4\pi V}{c^3} f^2 \tag{8.36}$$

(see Figure 8.7). Combined with the 3-dB bandwidth of the modes,

$$\Delta f = \frac{1}{2\pi \tau_m} = \frac{3 \ln 10}{\pi T_{\text{rev}}} \simeq \frac{cA}{8\pi V}, \tag{8.37}$$

where T_{rev} is the reverberation time[5] and A is the total absorption area of the room, this leads to the following expression for the modal overlap,

$$M = n(f)\Delta f \simeq \frac{A}{2c^2} f^2 \simeq \frac{12 \ln 10 \, V}{T_{\text{rev}} c^3} f^2. \tag{8.38}$$

The modal overlap is the average number of modes excited by a pure tone under the simplifying assumption that each mode is excited only if the frequency of the pure tone is within a band of the width Δf centred at the natural frequency of the mode.[6] The frequency

$$f_s = \sqrt{\frac{c^3 T_{\text{rev}}}{4 \ln 10 \cdot V}} \tag{8.39}$$

---

[5] As shown in Section 8.5 the reverberation time is proportional to the time constant: $T_{\text{rev}} = 6 \ln(10)\tau \simeq 13.8\tau$.
[6] Some authors prefer a slightly different measure, the statistical modal overlap, $M_s = \pi M$, which is based on the statistical modal bandwidth.

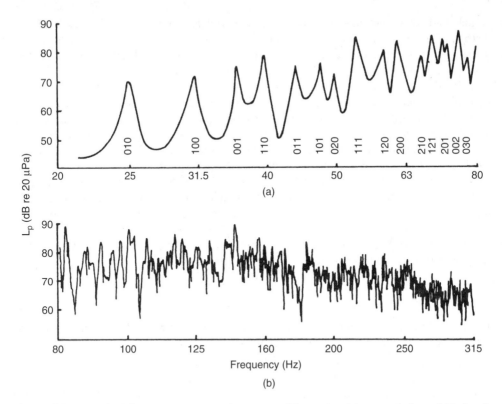

Figure 8.7 A typical frequency response in a room [7], reprinted by permission of Taylor & Francis

(usually written as $2000\sqrt{T_{rev}/V}$ with T_{rev} in seconds and V in cubic metres) is known as the Schroeder frequency or 'Schroeder's large room frequency'.[7] The Schroeder frequency is the frequency above which there is sufficient modal overlap to justify a statistical approach (via the central-limit theorem) even for pure-tone excitation.[8] Above this frequency the modal overlap exceeds (approximately) three, and experience shows that this is sufficient to justify the statistical approach [9].[9]

8.3.1 The Perfectly Diffuse Sound Field

A particularly simple statistical model of the sound field in a reverberation room assumes that the sound field is 'perfectly diffuse'. The perfectly diffuse sound field is composed of sound waves coming from all directions. This leads to the concept of a sound field in

---

[7] This concept is named after M.R. Schroeder, who derived fundamental parts of the statistical theory in a paper published in 1954 [8].
[8] Note that the losses of a reverberation room and thus T_{rev} in general depend on the frequency, which means that it may be necessary to calculate f_s at different frequencies. However, the reverberation time is usually a slowly varying function of frequency. It is often measured in one-third octave bands.
[9] With a modal overlap of three, the sound field in a rectangular room is essentially composed of 24 plane waves (cf. Equation (8.11)).

an unbounded medium generated by distant, uncorrelated sources of random noise evenly distributed over all directions. Since the sources are uncorrelated there would not be interference phenomena in such a sound field, and the field would therefore be completely homogeneous and isotropic. For example, the sound pressure level would be the same at all positions, and temporal correlation functions between linear quantities measured at two points would depend only on the distance between the two points. The time-averaged sound intensity would be zero at all positions. An approximation to this perfectly diffuse sound field might be generated by a number of loudspeakers driven with uncorrelated noise in a large anechoic room. The sound field in a reverberation room driven by one single source is quite different, of course. Nevertheless, combined with simple energy balance considerations the assumption that the sound field in a reverberation room is perfectly diffuse leads to an extremely useful relation between the sound power emitted by a source of noise in the room, the total absorption of the room, and the sound pressure generated by the source.

The derivation is fairly simple. It is assumed that the sound field in a reverberation room is composed of incoherent plane waves of the form

$$\hat{p}(\theta, \varphi) = \frac{C}{\sqrt{4\pi}} e^{j(\omega t - k_x x - k_y y - k_z z)}.$$
(8.40)

All waves have an amplitude of $C/\sqrt{4\pi}$, and they arrive from all directions, given by the three components of the wavenumber vector,

$$k_x = k \sin\theta \cos\varphi, \quad k_y = k \sin\theta \sin\varphi, \quad k_z = k \cos\theta.$$
(8.41a, 8.41b, 8.41c)

This model is based on the assumption that interference between the various waves can be ignored, which leads to a uniform mean-square pressure of

$$p_{\text{rms}}^2 = \frac{|C|^2}{2 \cdot 4\pi} \int_{-\pi}^{\pi} d\varphi \int_0^{\pi} \sin\theta \, d\theta = \frac{|C|^2}{2},$$
(8.42)

except, as we shall see later, near the walls. The z-component of the particle velocity in a single wave is

$$\hat{u}_z(\theta, \varphi) = \frac{C}{\rho c \sqrt{4\pi}} \frac{k_z}{k} e^{j(\omega t - k_x x - k_y y - k_z z)} = \frac{C \cos\theta}{\rho c \sqrt{4\pi}} e^{j(\omega t - k_x x - k_y y - k_z z)},$$
(8.43)

and the corresponding sound intensity is

$$I_z(\theta, \varphi) = \frac{1}{2} \text{Re} \left\{ \hat{p}(\theta, \varphi) \hat{u}_z^*(\theta, \varphi) \right\} = \frac{|C|^2}{8\pi\rho c} \cos\theta.$$
(8.44)

If this is integrated over the full solid angle of 4π the result is zero; there is no net intensity in any direction in the perfectly diffuse sound field. However, the boundaries of the room are exposed to a finite *incident* sound intensity. For example, the sound power incident per unit area on a wall at $z = l_z$ is found by restricting the sound incidence to a hemisphere,

$$I_{\text{inc},z} = \frac{|C|^2}{8\pi\rho c} \int_{-\pi}^{\pi} d\varphi \int_0^{\pi/2} \cos\theta \sin\theta \, d\theta = \frac{|C|^2}{8\rho c} = \frac{p_{\text{rms}}^2}{4\rho c},$$
(8.45)

Note that this is four times less than in a plane wave of normal incidence.

The losses of the walls are described in terms of the absorption coefficient (or absorption factor) of the walls, α. This is the absorbed fraction of the incident sound power.[10] Obviously, the absorption coefficient of any material must take values between naught and unity ($0 \leq \alpha \leq 1$). The total absorption of the room, A, is calculated by multiplying each area with its absorption coefficient,

$$A = \sum_i S_i \alpha_i. \tag{8.46}$$

Note that the unit of the room absorption A is m^2.

In the steady state the sound power emitted by the source is counterbalanced by the sound power absorbed by the walls. The latter is simply the product of the incident sound power per unit area and the total absorption, so we finally obtain the equation

$$P_{a,\text{source}} = P_{a,\text{abs}} = I_{\text{inc}} A = \frac{p_{\text{rms}}^2}{4\rho c} A. \tag{8.47}$$

Equation (8.47) makes it possible to predict the sound pressure generated by a source of known sound power or estimate the sound power of a source from the sound pressure it generates in a reverberation room. All that is required is the total absorption of the room. A more accurate version of this equation is derived in Section 8.5.1.

The theory of the perfectly diffuse sound field assumes that the waves that constitute the sound field are uncorrelated. Therefore interference effects between the waves are ignored, and the mean square pressures of the various waves are regarded as additive. However, it is possible to extend the theory to take account of the interference phenomena that occur near the walls of the room. The theory is due to Waterhouse [10]. Waterhouse assumed that an infinitely large, perfectly rigid plane surface was exposed to perfectly diffuse sound incidence.

Near the wall at $z = 0$ Equation (8.40) should be modified to

$$\hat{p}(z, \theta, \varphi) = \frac{C}{\sqrt{4\pi}}(e^{-jk_z z} + e^{jk_z z})e^{j(\omega t - k_x x - k_y y)} = \frac{C}{\sqrt{\pi}} \cos k_z z \, e^{j(\omega t - k_x x - k_y y)}, \tag{8.48}$$

because each incident wave will be fully coherent with the corresponding reflected wave. The sound incidence is now restricted to a hemisphere, and Equation (8.42) becomes

$$p_{\text{rms}}^2(z) = \frac{|C|^2}{2\pi} \int_{-\pi}^{\pi} d\varphi \int_0^{\pi/2} \cos^2(kz \cos \theta) \sin \theta \, d\theta$$

$$= \frac{|C|^2}{2} \int_0^{\pi/2} (1 + \cos(2kz \cos \theta)) \sin \theta \, d\theta = \frac{|C|^2}{2} \left(1 + \frac{\sin 2kz}{2kz}\right), \tag{8.49}$$

which shows that the interference between each wave and its reflected counterpart gives rise to an interference pattern near the boundaries of the room; see Figure 8.8. Note that the mean square sound pressure is doubled at the surface of the wall, corresponding to an increase of the level of 3 dB. Similar considerations show that there is a systematic increase of the sound pressure level of 6 dB near the edges of a reverberation room and an increase of 9 dB near the corners of the room [10].

---

[10] In general the absorption coefficient of a given material depends on the nature of the sound field. Here we are concerned with the diffuse field (or random incidence) absorption coefficient.

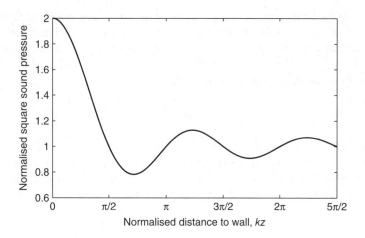

Figure 8.8   Interference pattern in a perfectly diffuse sound field in front of a rigid wall

Waterhouse used Equation (8.49) to calculate the fractional increase of the total sound energy in the room due to the interference pattern near the walls [10]. Ignoring the interference effects near edges and corners he obtained the following approximate expression

$$\frac{S}{V}\int_0^\infty \frac{\sin(2kh)}{2kh}\,\mathrm{d}h = \frac{S}{2kV}\int_0^\infty \frac{\sin x}{x}\,\mathrm{d}x = \frac{\pi S}{4kV} = \frac{S\lambda}{8V}, \tag{8.50}$$

where λ is the wavelength, and concluded that, instead of the total sound energy $V p_{\mathrm{rms}}^2/(\rho c^2)$, which is what one would expect,[11] the total sound energy in the room amounts to

$$E_{\mathrm{a}} \simeq V \frac{p_{\mathrm{rms}}^2}{\rho c^2}\left(1 + \frac{S\lambda}{8V}\right), \tag{8.51}$$

where p_{rms} should be interpreted as the sound pressure measured in the central region of the room. The factor in parenthesis is known as the 'Waterhouse correction'. It is negligibly small except at low frequencies. In Section 8.5.1 we shall use this quantity for improving Equation (8.47).

8.3.2   The Sound Field in a Reverberation Room Driven with a Pure Tone

Exceedingly useful as it is, the perfectly diffuse field model is too coarse an approximation for some purposes. For example, it ignores the interference phenomena that are caused by the fact all the sound waves in the room are generated by the same source and postulates that the sound field is homogeneous (in obvious disagreement with the pure-tone expressions used in deriving Equations (8.40) and (8.43), and this is not a good

---

[11] In a resonant system the time averaged potential energy equals the time averaged kinetic energy. Therefore the total energy is simply twice the potential energy.

approximation in a reverberation room driven with a narrow-band signal. A more realistic model of the sound field in a reverberation room above the Schroeder frequency describes the sound field as composed of *coherent* plane waves with random phases arriving from all directions. This is a pure-tone model, and therefore the various plane waves interfere in the entire sound field, not just near the boundaries as assumed in Section 8.3.1. As we shall see, the result is a sound field in which the sound pressure level depends on position, although the *probability* of the level being in a certain interval is the same at all positions. Temporal correlation functions between linear quantities measured at two positions depend on the positions, although the *probability* of a given correlation function being in a certain interval depends only on the distance between the two points. The time-averaged sound intensity assumes a finite value at all positions. Since infinitely many plane waves with completely random phases are assumed, this model is also idealised, but it gives a good approximation to the sound field in a reverberation room driven with a pure tone above the Schroeder frequency. With averaging over an ensemble of realisations the perfectly diffuse field described above is obtained. An approximation to ensemble averaging is obtained if the room is equipped with a rotating diffuser.[12]

As mentioned above, the statistical model is based on the assumption that the sound field can be modelled as a sum of plane waves with random phases and amplitudes, arriving from random directions. Therefore the sound pressure at a given position can be written

$$\hat{p}(\mathbf{r}_1) = \lim_{N \to \infty} \frac{1}{\sqrt{N}} \sum_{i=1}^{N} A_i e^{j(\omega t - \mathbf{k}_i \cdot \mathbf{r}_1)} = \lim_{N \to \infty} \frac{1}{\sqrt{N}} \sum_{i=1}^{N} A_i e^{j(\omega t - k_x x_1 - k_y y_1 - k_z z_1)}, \quad (8.52)$$

where A_i is the complex amplitude of the i'th wave and $\mathbf{k}_i$ is its wavenumber vector, that is, a vector of the length k ($= \omega/c$) pointing in the direction of propagation of the i'th wave (cf. Equation (8.41)). The amplitudes are independent random variables with uniformly distributed phases, and all directions of the wavenumber vectors are equally probable. At another position in the sound field the sound pressure is

$$\hat{p}(\mathbf{r}_2) = \lim_{N \to \infty} \frac{1}{\sqrt{N}} \sum_{i=1}^{N} A_i e^{j(\omega t - \mathbf{k}_i \cdot \mathbf{r}_2)} = \lim_{N \to \infty} \frac{1}{\sqrt{N}} \sum_{i=1}^{N} A_i e^{j(\omega t - \mathbf{k}_i \cdot \mathbf{r}_1)} e^{j\mathbf{k}_i \cdot \mathbf{r}}, \quad (8.53)$$

where $\mathbf{r}$ is the vector that separates the two points. Thus $\mathbf{k}_i \cdot \mathbf{r}$ is a term that accounts for the phase shift between point no 1 and point no 2. Since the directions of the various waves are random this is a random term, and if $r > \lambda/2$ (where $r = |\mathbf{r}|$) it takes random values over an interval of more than $\pm\pi$. In other words, at points more than half a wavelength apart we have in effect independent sets of random phases. This leads to the conclusion that the statistical properties of the sound field with respect to position can be determined by studying the *ensemble statistics* of sums of the type given by Equation (8.52).

The spatial statistics of the mean square pressure is particularly interesting, since spatial average values of this quantity are central in practically all measurements in reverberation rooms. The amount of spatial averaging that is required to ensure a reliable estimate is

---

[12] A rotating diffuser is a large, slowly rotating device that changes the modal pattern in the room.

obviously of concern. The mean square pressure can be written

$$
p_{\text{rms}}^2(\mathbf{r}_1) = \lim_{N \to \infty} \frac{1}{2N} \left| \sum_{i=1}^{N} A_i e^{j(\omega t - \mathbf{k}_i \cdot \mathbf{r}_1)} \right|^2
$$

$$
= \lim_{N \to \infty} \frac{1}{2N} \left| \sum_{i=1}^{N} |A_i| \left(\cos(\varphi_i - \mathbf{k}_i \cdot \mathbf{r}_1) + j \sin(\varphi_i - \mathbf{k}_i \cdot \mathbf{r}_1) \right) \right|^2
$$

$$
= \lim_{N \to \infty} \frac{1}{2N} \left(\left(\sum_{i=1}^{N} |A_i| \cos(\varphi_i - \mathbf{k}_i \cdot \mathbf{r}_1) \right)^2 + \left(\sum_{i=1}^{N} |A_i| \sin(\varphi_i - \mathbf{k}_i \cdot \mathbf{r}_1) \right)^2 \right),
$$

$$(8.54)$$

where φ_i is the phase of A_i. It can be seen that the mean square pressure is a sum of two squared sums of random terms. This is a random variable, and from the central limit theorem we conclude that each sum is a Gaussian variable (see Appendix D). It can readily be shown that the two sums have the same statistical properties, and that they are independent random variables. A sum of n squared, independent Gaussian random variables with the same distribution has a chi-square distribution with n degrees of freedom (see Appendix D). In the particular case of $n = 2$ this is the *exponential distribution*, which leads to the conclusion that the mean square pressure in the room is a random variable with the probability density function

$$
f_{p_{\text{rms}}^2}(x) = \begin{cases} \dfrac{1}{\xi} e^{-x/\xi} & \text{if } x > 0 \\ 0 & \text{elsewhere,} \end{cases}
$$

$$(8.55)$$

where

$$
\xi = E\{p_{\text{rms}}^2\}
$$

$$(8.56)$$

is the ensemble average of the mean square pressure, which is also the spatial average value. The exponential probability density is shown in Figure 8.9. Equation (8.55) shows that the probability of the mean square pressure at a given position in the room exceeding a certain value ε can be written

$$
P\{p_{\text{rms}}^2 > \varepsilon\} = \int_{\varepsilon}^{\infty} f_{p_{\text{rms}}^2}(x) \mathrm{d}x = \frac{1}{E\{p_{\text{rms}}^2\}} \int_{\varepsilon}^{\infty} \exp(-x/E\{p_{\text{rms}}^2\}) \mathrm{d}x.
$$

$$(8.57)$$

One of the properties of an exponentially distributed random variable is that its relative standard deviation, that is, the standard deviation divided by the average value, equals unity; see Appendix D. From the foregoing we can now conclude that the mean square pressure in a reverberation room driven with a pure tone varies significantly with the position: its relative spatial standard deviation equals unity.

It can be shown from Equation (8.55) that the probability density of the sound pressure *level* is [8]

$$
f_{L_p}(x) = \frac{\ln 10}{10} \exp\left(\frac{\ln 10}{10} (x - L_0) - \exp\left(\frac{\ln 10}{10} (x - L_0) \right) \right),
$$

$$(8.58)$$

where L_0 is the level that corresponds to $E\{p_{rms}^2\}$. This function is shown in Figure 8.10. The corresponding standard deviation is about 5.6 dB. See also Figure 8.11, which shows the sound pressure level recorded by a traversing microphone in a reverberation room.

Finally it is worth calling attention to the fact that no properties of the room have entered into these considerations – except a modal overlap of at least three. According to this theory the spatial standard deviation of the sound pressure level is about 5.6 dB in *any* lightly damped room driven with a pure tone above the Schroeder frequency except close to the source that generates the sound field.

It is physically obvious that the sound pressures at two points very close to each other tend to be similar; there is a limit to how rapidly the sound field changes with position.

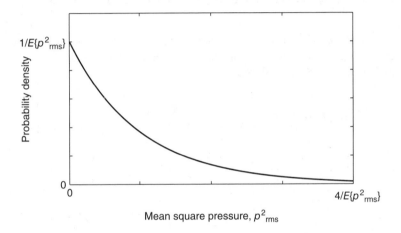

Figure 8.9   The exponential distribution

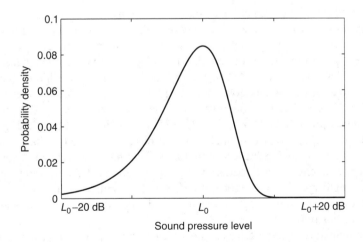

Figure 8.10   The probability density of the sound pressure level in a reverberation room driven with a pure tone

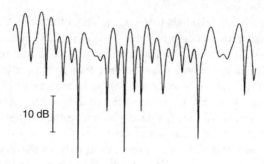

10 dB

Figure 8.11   The sound pressure level as a function of position on a straight line in a reverberation room driven with a pure tone [11], reprinted by permission of Taylor & Francis

A statistical analysis of such a phenomenon leads to a description in terms of correlation functions. We can calculate the spatial correlation of the sound pressure as follows. Combining Equations (8.52) and (8.53) gives

$$E\{\hat{p}(\mathbf{r}_1)\hat{p}^*(\mathbf{r}_2)\} = E\left\{\lim_{N\to\infty}\frac{1}{N}\sum_i^N\sum_j^N |A_i A_j|\, e^{j(\varphi_i-\varphi_j)}e^{-j\mathbf{k}_i\cdot\mathbf{r}}\right\}$$

$$= E\left\{\lim_{N\to\infty}\frac{1}{N}\sum_i |A_i|^2 \cos(\mathbf{k}_i\cdot\mathbf{r})\right\}$$

$$= E\{|p|^2\}\frac{1}{4\pi}\int_0^{2\pi}\int_0^\pi \cos(kr\cos\theta)\sin\theta\, d\theta\, d\varphi$$

$$= E\{|p|^2\}\frac{1}{2kr}\int_{-kr}^{kr}\cos x\, dx = E\{|p|^2\}\frac{\sin kr}{kr}. \tag{8.59}$$

The normalised correlation function is shown in Figure 8.12. The interpretation of this spatial correlation function is that sound pressure signals recorded at positions less than, say, a quarter of a wavelength apart in a reverberation room driven with a pure tone are likely to have almost the same amplitude and phase. On the other hand, if there is more than half a wavelength between the two points then knowledge of the amplitude and phase of one of the signals gives practically no knowledge of the other signal.

Using similar considerations one can show that the normalised spatial covariance of the mean square pressure is $(\sin kr/kr)^2$ [12]. This function is shown in Figure 8.13. It is apparent that the covariance is negligible for distances exceeding half a wavelength, indicating that microphone positions should be spaced half a wavelength apart (or more) for maximum efficiency of the spatial averaging procedure. If the positions are less than half a wavelength apart they do not give independent sample values. With independent sample values the uncertainty of the resulting estimate is reduced by the square root of the number of positions.

Various continuous averaging procedures have been examined in [13, 14]. One of the most important results is that averaging over a straight path of the length l is equivalent to averaging over $l/(\lambda/2)$ independent positions. In practice the spatial averaging is often

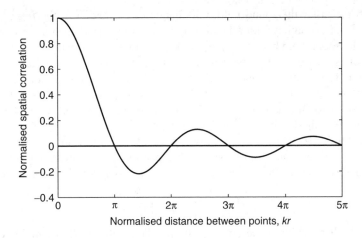

Figure 8.12   The spatial correlation of the sound pressure in a reverberation room driven with a pure tone

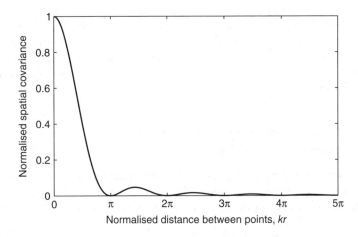

Figure 8.13   The spatial covariance of the mean square sound pressure in a reverberation room driven with a pure tone

carried out using a rotating microphone boom. If the diameter of the measurement path exceeds one wavelength then the equivalent number of independent positions may be calculated as if it were straight.

8.3.3   Frequency Averaging

It is possible to extend the statistical pure-tone model to excitation with a band of noise. If a reverberation room is driven with a pure tone whose frequency is shifted slightly, the phases and amplitudes of the plane waves that compose the sound field are changed, which means that the entire interference pattern is changed [8]. The longer the reverberation time of the room the faster the sound field will change as a function of the frequency [15]. It now follows that excitation with a band of noise corresponds to averaging over the band. As a result the sound field becomes more uniform, temporal correlation functions between linear quantities measured at pairs of positions tend to depend less on the particular positions, the sound intensity is reduced, and the sound power of a monopole approaches the free field power. The effect of this spectral averaging depends not only on the bandwidth of the excitation (or the analysis) but also on the damping of the room; the longer the reverberation time the more efficient the averaging.

It can be shown [16, 17] that the relative spatial standard deviation of the mean square pressure in a reverberation room driven with a band of noise (or driven with wide band noise and analysed in bands) to a good approximation is given by the expression

$$\varepsilon\{p_{rms}^2\} \simeq \frac{1}{\sqrt{1 + (2B\tau)}}, \tag{8.60}$$

where B is the bandwidth of the noise (in hertz). This will usually be much less than the relative standard deviation of unity found in a room driven with a pure tone.

It is apparent that there are many similarities between the sound field in a room driven with noise and the perfectly diffuse sound field described in Section 8.3.1; this explains the usefulness of this concept. However, there are also important differences between the 'fine structure' of the sound field in a real room and a perfectly diffuse sound field; this can be observed when the sound field is analysed with fine spectral resolution. It can also be concluded that diffuseness at low frequencies requires a large room, and that long reverberation times (within the limits determined by the requirement of sufficient modal overlap) are favourable.

8.3.4   The Sound Power Emitted by a Point Source
in a Lightly Damped Room

The sound power emitted by a point source in a lightly damped room cannot be expected to be unaffected by the reverberant sound field even when it is placed far from the walls of the room. The sound power of a pure-tone monopole with the volume velocity $Q\mathrm{e}^{j\omega t}$ is, as we shall see in Section 9.2,

$$P_a = \frac{|Q|^2}{2}\mathrm{Re}\{Z_r\} \tag{8.61}$$

where Z_r is the radiation impedance (cf. Equation (6.13)). The sound field at the source position may be regarded as the sum of the direct field and the reverberant field, and therefore the radiation impedance is the sum of the free-field radiation impedance and the complex ratio of the sound pressure associated with the reverberant field and the volume velocity of the source,

$$Z_r = \frac{\rho c k^2}{4\pi} + \frac{\hat{p}_{rev}}{Q\mathrm{e}^{j\omega t}}. \tag{8.62}$$

All phases are equally probable in the reverberant field, and thus $E\{\hat{p}_{rev}\} = 0$, which leads to the conclusion that *on average* the monopole emits its free-field sound power output

$$E\{P_a\} = \frac{\rho c k^2 |Q|^2}{8\pi}. \tag{8.63}$$

However, the actual sound power output of the source varies with the position. The corresponding spatial variance can be calculated as follows,

$$
\sigma^2 \{P_a\} = \frac{|Q|^4}{4} \sigma^2 \{\mathrm{Re}\,\{Z_r\}\} = \frac{|Q|^4}{4} \sigma^2 \left\{\mathrm{Re}\left\{\frac{\hat{p}_{\mathrm{rev}}}{Q e^{j\omega t}}\right\}\right\} = \frac{|Q|^2}{4} E\left\{|\hat{p}_{\mathrm{rev}}|^2 \cos^2(\varphi)\right\}
$$

$$
= \frac{|Q|^2}{8} E\left\{|\hat{p}_{\mathrm{rev}}|^2\right\} \simeq \frac{|Q|^2}{8} \frac{8\rho c}{A} E\{P_a\} \simeq E^2\{P_a\} \frac{8\pi}{k^2 A}, \tag{8.64}
$$

where use has been made of Equations (8.47) and (8.63) and the fact that the average of a squared cosine of a random phase is $1/2$. However, this derivation has not taken account of the phenomenon known as 'weak Anderson localisation' (also known as 'coherent backscattering'), according to which there is a concentration of the *reverberant* part of the sound field exactly at the source position [18]. This effect increases the relative variance by a factor of 2. It can now be seen that the relative standard deviation is inversely proportional to the modal overlap,

$$
\varepsilon\{P_a\} \sim \frac{1}{k}\sqrt{\frac{16\pi}{A}} = \sqrt{\frac{2}{M\pi}} \tag{8.65}
$$

(cf. Equation (8.38)). The resulting relation is compared with experimental results in Figure 8.14.

It is apparent that the sound power emitted by the source at low frequencies varies substantially with room and position unless the room is very large and heavily damped. This indicates that it is extremely important to average over many source positions if the sound power of a source with a significant content of pure tones is to be determined in a reverberation room. Ideally one should also average over several rooms.

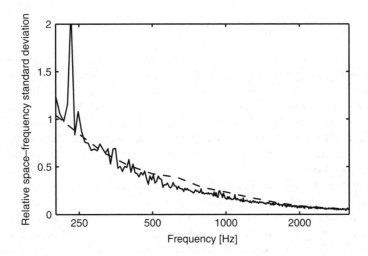

Figure 8.14 Predicted and measured relative ensemble standard deviation of the sound power output of a monopole that emits a pure tone (Equation (8.65))

If the source emits bandpass filtered noise the spatial standard deviation is reduced by the factor given by Equation (8.60). Thus

$$\varepsilon\{P_a\} \simeq \frac{1}{k_0}\sqrt{\frac{16\pi}{A}}\frac{1}{\sqrt{1+(2B\tau)}} = \frac{1}{k_0}\sqrt{\frac{16\pi}{A}}\frac{1}{\sqrt{1+8BV/cA}} \simeq \frac{1}{k_0}\sqrt{\frac{2\pi c}{BV}}, \qquad (8.66)$$

where k_0 corresponds to the centre frequency. In this case the damping of the room has almost no influence. It can be seen that although the relative standard deviation of the emitted sound power is significantly reduced by the frequency averaging it is still not negligible at low frequencies unless the room is very large. It must be concluded that there is a considerable uncertainty in sound power measurements and in predictions based on sound power at low frequencies.

Extending the random wave model to the region below the Schroeder frequency

The model presented in Section 8.3.2 is only valid when the modal overlap of the room is high. In this frequency range the source emits essentially its free field sound power, and spatial statistics corresponds to ensemble statistics. However, it is possible to extend the model to lower frequencies by taking account of the variations of the sound power of the source. These random variations modify the ensemble average of the amplitudes of the waves. The result is an additional contribution to the variance of the mean square pressure that is inversely proportional to the modal overlap [19]; and spatial statistics is no longer the same as ensemble statistics, since it matters whether the frequency coincides with a natural frequency or not.

Statistical results based on the modal theory

It is possible to derive expressions in closed form for statistical properties of the sound power output of the source and the mean square pressure in the room by averaging expressions determined from Equation (8.35) over source and/or receiver position and replacing the resulting modal sums by integrals [20–22]. The latter procedure may be interpreted as determining averages over an ensemble of rooms with slightly different dimensions but essentially the same size. Such rooms would have the same modal density, but the particular distributions of modal frequencies would be different.

The theory is fairly complicated because the modal frequencies tend to exhibit 'long range repulsion', and thus they are not independently distributed on the frequency axis with a density given by Equation (8.15) as originally though; they are distributed in accordance with the random matrix theory of Gaussian orthogonal ensembles [23]. This theory is well established, and its results are in good agreement with the recently extended random wave theory; see [19].

8.4   The Decay of Sound in a Lightly Damped Room

Consider a room driven by a source that emits sound, but not necessarily steady sound. Simple energy balance considerations lead to the following equation,

$$P_{\text{a,source}} - P_{\text{a,abs}} = \frac{\mathrm{d}E_a}{\mathrm{d}t}, \tag{8.67}$$

where $P_{\text{a,source}}$ is the sound power emitted by the source, $P_{\text{a,abs}}$ is the sound power absorbed by the walls of the room, and E_a is the total sound energy in the room. (Obviously, Equation (8.67) is an extension of Equation (8.47).) If the source is suddenly turned off Equation (8.67) becomes

$$P_{\text{a,abs}} + \frac{\mathrm{d}E_a}{\mathrm{d}t} = 0, \tag{8.68}$$

which is the basis for the following simple considerations. We will study the problem using the two fundamentally different models presented in Sections 8.2 and 8.3.

8.4.1   The Modal Approach to Decay of Sound

In this section we will study the sound field mode by mode. The walls of the room have a finite but small admittance with a finite real part, indicating that acoustic energy is absorbed by the walls. The following considerations are based on the assumption that the mode shapes in the lightly damped room are the same as in the undamped room. It is also assumed that each mode maintains its shape during the decay process; it is just the amplitude that decreases with time.

We can express the total sound energy in the mode in terms of an integral of the squared mode shape as follows,

$$E_{\text{a},m} = \int_V (w_{\text{kin},m} + w_{\text{pot},m})\mathrm{d}V = \int_V 2w_{\text{pot},m}\mathrm{d}V = \int_V \frac{|\hat{p}_m|^2}{2\rho c^2}\mathrm{d}V = \int_V \frac{A_m^2 \psi_m^2}{2\rho c^2}\mathrm{d}V, \tag{8.69}$$

where A_m is the amplitude of the mode (cf. Equation (8.7)). The sound power absorbed by the walls is obtained by integrating the product of the mean square sound pressure and the real part of the local admittance over the surface that defines the room,

$$P_{\text{a,abs},m} = \int_S \mathbf{I} \cdot \mathrm{d}\mathbf{S} = \int_S \frac{|\hat{p}_m|^2}{2}\mathrm{Re}\{Y_\text{s}\}\mathrm{d}S = \int_S \frac{A_m^2 \psi_m^2}{2}\mathrm{Re}\{Y_\text{s}\}\mathrm{d}S, \tag{8.70}$$

where Y_s is the local specific wall admittance (the local ratio of the normal component of the particle velocity to the sound pressure). This is the reciprocal of the local specific wall impedance. Equation (8.70) is based on the assumption that the walls are locally reacting, which implies that Y is independent of the direction of sound incidence on the surface.

Under the reasonable assumption that the mode shape is maintained during the decay process we can expect the ratio $E_a/P_{\text{a,abs}}$ associated with a given mode to be constant. Apparently this ratio has the dimension of time,

$$\tau_m = \frac{E_{\text{a},m}}{P_{\text{a,abs},m}} = \frac{1}{\rho c^2} \frac{\displaystyle\int_V \psi_m^2 \mathrm{d}V}{\displaystyle\int_S \psi_m^2 \mathrm{Re}\{Y_\text{s}\}\mathrm{d}S}. \tag{8.71}$$

We can now write Equation (8.68) in the form

$$\frac{E_{a,m}}{\tau_m} + \frac{dE_{a,m}}{dt} = 0. \tag{8.72}$$

The solution to this simple first-order differential equation is a decaying exponential,

$$E_{a,m}(t) = E_{a,m}(t_0)e^{-(t-t_0)/\tau_m} \quad \text{for } t \geq t_0, \tag{8.73}$$

where t_0 is the time at which the source is switched off. We conclude that the sound energy in each mode decreases exponentially with a time constant given by Equation (8.71).

The decay constants of the modes in a rectangular room with uniform wall admittance can be found by combining Equations (8.6) and (8.71). Because of the normalisation, the volume integral in the numerator of Equation (8.70) simply equals V (cf. Equation (8.23)). The surface integral of the denominator becomes

$$\int_S \psi_m^2 \mathrm{Re}\{Y_s\} dS \simeq \begin{cases} 8\mathrm{Re}\left\{Y_s\right\} S\dfrac{1}{4} & \text{for oblique modes,} \\[2ex] 4\mathrm{Re}\{Y_s\} S \left(\dfrac{1}{3}\dfrac{1}{4} + \dfrac{2}{3}\dfrac{1}{2}\right) & \text{for tangential modes,} \\[2ex] 2\mathrm{Re}\{Y_s\} S \left(\dfrac{1}{3} + \dfrac{2}{3}\dfrac{1}{2}\right) & \text{for axial modes,} \end{cases} \tag{8.74}$$

where the first factor (8, 4 or 2) is the square of the normalisation constant. The equation is exact for oblique modes; for axial and tangential modes it has been assumed that $l_x \simeq l_y \simeq l_z$ so that each wall has an area of $S/6$. Inserting in Equation (8.71) gives

$$\tau_m \simeq \begin{cases} \dfrac{V}{2\rho c^2 S\mathrm{Re}\left\{Y_s\right\}} = \dfrac{V}{2cS\,\mathrm{Re}\{\beta\}} & \text{for oblique modes,} \\[2ex] \dfrac{3V}{5\rho c^2 S\mathrm{Re}\{Y_s\}} = \dfrac{3V}{5cS\,\mathrm{Re}\{\beta\}} & \text{for tangential modes,} \\[2ex] \dfrac{3V}{4\rho c^2 S\mathrm{Re}\{Y_s\}} = \dfrac{3V}{4cS\,\mathrm{Re}\{\beta\}} & \text{for axial modes,} \end{cases} \tag{8.75}$$

where for simplicity we have introduced the normalised (dimensionless) specific wall admittance

$$\beta = \rho c Y_s. \tag{8.76}$$

It is apparent that axial modes die out at a slightly slower rate than tangential and oblique modes. In practice the losses of rooms are measured in frequency bands, say, one-third octave bands, which means that many modes are excited at the same time. As a result of the different decay rates logarithmic decay functions recorded in rectangular rooms without diffusing plates[13] tend to be curved at low frequencies; see Figure 8.15. At higher

---

[13] Diffusers in the form of large stationary plates distributed at random in the room and having random orientation tend to couple the various modes and thus make the decay curves more linear.

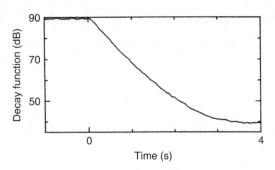

Figure 8.15   A curved decay curve recorded in a reverberation room [24], reprinted by permission of Multi-Science Publishing

frequencies oblique modes dominate, and the logarithmic decay functions tend to be almost perfectly linear.

If we consider a rectangular room in which the losses are concentrated on one of the walls, say $z = 0$, a similar analysis gives

$$\int_S \psi_m^2 \mathrm{Re}\{Y_s\} \mathrm{d}S = \begin{cases} 8\mathrm{Re}\{Y_s\} l_x l_y \dfrac{1}{4} & \text{for oblique modes,} \\[2mm] 4\mathrm{Re}\{Y_s\} l_x l_y \dfrac{1}{4} & \text{if } n_x > 0 \text{ and } n_y > 0 \text{ and } n_z = 0, \\[2mm] 4\mathrm{Re}\{Y_s\} l_x l_y \dfrac{1}{2} & \text{if } n_x = 0 \text{ and } n_y > 0 \text{ and } n_z > 0, \\[2mm] 4\mathrm{Re}\{Y_s\} l_x l_y \dfrac{1}{2} & \text{if } n_x > 0 \text{ and } n_y = 0 \text{ and } n_z > 0, \\[2mm] 2\mathrm{Re}\{Y_s\} l_x l_y & \text{if } n_x = 0 \text{ and } n_y = 0 \text{ and } n_z > 0, \\[2mm] 2\mathrm{Re}\{Y_s\} l_x l_y \dfrac{1}{2} & \text{if } n_x > 0 \text{ and } n_y = 0 \text{ and } n_z = 0, \\[2mm] 2\mathrm{Re}\{Y_s\} l_x l_y \dfrac{1}{2} & \text{if } n_x = 0 \text{ and } n_y > 0 \text{ and } n_z = 0, \end{cases} \qquad (8.77)$$

which leads to the conclusion that there are only two different values of τ_m, depending on whether there is wave motion in the z-direction or not,

$$\tau_m = \begin{cases} \dfrac{l_z}{2c\mathrm{Re}\{\beta\}} & \text{for } n_z \neq 0, \\[3mm] \dfrac{l_z}{c\mathrm{Re}\{\beta\}} & \text{for } n_z = 0. \end{cases} \qquad (8.78)$$

Note that the two decay constants differ by a factor of two, which is more than in the room with uniform absorption. In rooms with dominating absorption on just one surface the decay curves tend to be more curved than in rooms with a more uniform distribution of the losses.

8.4.2 The Statistical Approach to Decay of Sound

Assuming a stationary ideal diffuse sound field in the room we can express the total sound energy in the room as

$$E_a = V \frac{\langle p_{\text{rms}}^2 \rangle}{\rho c^2}, \tag{8.79}$$

where $\langle p_{\text{rms}}^2 \rangle$ is the spatial average of the mean square sound pressure.[14] The absorbed sound power is

$$P_{a,\text{abs}} = \int_S \mathbf{I} \cdot \mathbf{n} \, dS = \int_S I_{\text{inc}} \alpha \, dS = \frac{\langle p_{\text{rms}}^2 \rangle}{4\rho c} A \tag{8.80}$$

(cf. Equation (8.47)). Note that this is *not* a mode-by-mode analysis; by contrast we must assume excitation with a band of noise (cf. the considerations in Section 8.3.3). Under the reasonable assumption that the two expressions given by Equations (8.79) and (8.80) remain valid also during the decay process[15] we can now introduce the time constant τ, defined as the ratio

$$\tau = \frac{E_a}{P_{a,\text{abs}}} = \frac{4V}{cA}. \tag{8.81}$$

Assuming uniform absorption at the walls (which implies that $A = S\alpha$) and comparing Equation (8.75) (for three-dimensional modes) and Equation (8.81) leads to the conclusion that they agree if

$$\alpha = 8 \operatorname{Re}\{\beta\}. \tag{8.82}$$

To put this result in first order is in agreement with what is found by calculating the dissipated fraction of the incident sound power per unit area of an infinite plane surface with the normalised specific admittance β, assuming diffuse sound incidence (see, e.g., [2]).

If only the wall at $z = 0$ is absorbing Equation (8.80) becomes

$$P_{a,\text{abs}} = \int_S I_n \, dS = \frac{\langle p_{\text{rms}}^2 \rangle v_c}{4\rho c} l_x l_y \alpha, \tag{8.83}$$

and Equation (8.81) becomes

$$\tau = \frac{E_a}{P_{a,\text{abs}}} = \frac{4V}{c l_x l_y \alpha} = \frac{4 l_z}{c \alpha}, \tag{8.84}$$

and this is seen to be in agreement with Equation (8.78) (for oblique modes) for the same relation between α and $\operatorname{Re}\{\beta\}$, confirming the consistency of the modal and the statistical theory.

---

[14] One might introduce the Waterhouse correction in the expression for the sound energy, but for simplicity this is omitted since a comparable correction applies to the expression for the absorbed sound power [25]; and it is the ratio of these quantities that matters.

[15] Obviously p_{rms}^2 must be interpreted as a running short-time average of the squared sound pressure for this to be meaningful.

8.5   Applications of Reverberation Rooms

Reverberation rooms are used for a number of standardised measurements, some of the most important of which are briefly described in the following.

8.5.1   Sound Power Determination

Reverberation rooms are useful for measurement of sound power, in particular the sound power of machines that operate in long cycles, since the alternatives, the sound intensity method and the free field method, are not very suitable for such sources. The considerations presented in Sections 8.3.3 and 8.3.4 lead to the conclusions that to reduce the uncertainty associated with sound power measurements of sources that emit pure tones the reverberation room should be fairly damped, whereas the best facilities for measurements of the sound power of sources that essentially emit random noise are reverberation rooms with a long reverberation time. In both cases the rooms should be large.

In Section 8.3.1 a relation between the sound power emitted by a source in a reverberation room, the sound pressure generated by the source, and the total room absorption was derived (Equation (8.47)) on the basis of considerations that made use of the concept of a perfectly diffuse sound field. We should now be able to develop a more precise relation,

$$P_{\text{a,source}} = \frac{\langle p_{\text{rms}}^2 \rangle_{V_{\text{c}}} A}{4\rho c} \left(1 + \frac{S\lambda}{8V}\right) = \frac{\langle p_{\text{rms}}^2 \rangle_{V_{\text{c}}} V}{\rho c^2} \left(1 + \frac{S\lambda}{8V}\right) \frac{6\ln 10}{T_{\text{rev}}}, \tag{8.85}$$

where

$$T_{\text{rev}} = -60/d \tag{8.86}$$

is the reverberation time and d is the rate of decay in decibels per second,[16] here assumed constant.[17] Note that the Waterhouse correction (Equation (8.51)) has been applied to account for the additional sound energy near the boundaries of the room.[18]

Since, by definition,

$$10\log(e^{-T_{\text{rev}}/\tau}) = -60 \text{ (dB)}, \tag{8.87}$$

we have the following relation between the time constant and the reverberation time,

$$T_{\text{rev}} = 6\ln(10)\tau. \tag{8.88}$$

---

[16] It can happen that the rate of decay depends on the interval (cf. Figure 8.15). In such cases it is the initial rate of decay that matters, and in practice an interval of about 10 dB should be used in such cases, that is T_{rev} is the time it takes for the level to decrease by 10 dB multiplied by a factor of 6. Note, however, that narrow bandpass filters tend to distort the very first part of short acoustic decay functions [26, 27], so the interval should be chosen with care. At medium and high frequencies the logarithmic decay curves are usually almost perfectly linear, and then the interval does not matter.

[17] Since decays in rooms are not in general linear over a large dynamic range (cf. the considerations in Section 8.4) room acousticians concerned with perceived reverberance distinguish between the early decay time (often abbreviated EDT), T_{20} and T_{30}, determined respectively by fitting a straight line over an interval of 0 to -10 dB, -5 to -25 dB and -5 to -35 dB. In any case it is *the rate of decay* that is used for determining the time it takes for the straight line to decay 60 dB.

[18] The Waterhouse correction is negligible except at low frequencies. Since Waterhouse's derivation was based on an assumption of ideal diffuse sound incidence, one might expect the resulting 'correction' to be less accurate at low frequencies, which is where it really matters. However, much later numerical calculations have confirmed the validity of Equation (8.51) [28].

8.5.2   Measurement of Sound Absorption

Reverberation rooms are also used for measurement of the sound absorption of acoustic materials. In the standardised method of determining diffuse-field sound absorption a large sample of the material under test (10 m^2) is installed in the room, and the absorption area and thus the absorption coefficient is deduced from the resulting reduction of the reverberation time,

$$\alpha = \frac{24 \ln(10) V}{c S_s} \left(\frac{1}{T_{rev}^s} - \frac{1}{T_{rev}} \right), \tag{8.89}$$

where T_{rev}^s is the reverberation time with the absorbing specimen in the room, T_{rev} is the reverberation time without the material in the room, and S_s is the area of the specimen. However, it is well known that this method gives results that vary rather significantly from one room to another even under the conditions that are specified in the standard [29, 30], which shows that the simple statistical theory based on Equation (8.45) is, after all, only approximate.

8.5.3   Measurement of Transmission Loss

A suite of reverberation rooms can be used for measurement of the transmission loss of the partition between two adjoining rooms provided that (unwanted) flanking transmission is negligible. In the source room the sound power incident on the partition under test is deduced from the spatial average of the mean square pressure (cf. Equation (8.45)); and in the receiver room the transmitted sound power is determined as described in Section 8.5.1. The transmission loss is the ratio of incident to transmitted sound power (in decibels).

In the receiving room one should obviously use the Waterhouse correction, although this is not recommended by the measurement standard (for political reasons) [31]. In the source room one should use a similar correction, as shown only recently [25].

References

[1] H. Weyl: The asymptotic distribution law for the eigenvalues of linear partial differential equations. *Mathematical Annals* **71**, 441–479 (1912).

[2] P.M. Morse: *Vibration and Sound* (2nd ed.) The American Institute of Physics, New York, 1981. See Chapter 32.

[3] L. Cremer and H. Müller: *Principles and Applications of Room Acoustics* (Volume 2). Applied Science Publishers Ltd, London (1982). See Chapter 7.

[4] S.H. Crandall: The role of damping in vibration theory. *Journal of Sound and Vibration* **11**, 3–18 (1970).

[5] A.D. Pierce: *Acoustics. An Introduction to Its Physical Principles and Applications*. The American Institute of Physics, New York, 1989. See Chapter 6.

[6] K.J. Ebeling: Statistical properties of random wave fields. Chapter 4 in *Physical Acoustics, Principles and Methods* Vol. XVII, ed. W. P. Mason and R. N. Thurston. Academic Press, New York (1984).

[7] D.A. Bies and C.H. Hansen: *Engineering Noise Control* (2nd ed.). E & FN Spon, London (1996).

[8] M. Schröder: Die statistischen Parameter der Frequenzkurven von großen Räumen. *Acustica* **4**, 594–600 (1954). [An English translation, 'Statistical parameters of the frequency response curves of large rooms', has been published in *Journal of the Audio Engineering Society* **35**, 299–305 (1987).]

[9] M.R. Schroeder and K.H. Kuttruff: On frequency response curves in rooms. Comparisons of experimental, theoretical and Monte Carlo results for the average spacing between maxima. *Journal of the Acoustical Society of America* **34**, 76–80 (1962).

[10] R.V. Waterhouse: Interference patterns in reverberant sound fields. *Journal of the Acoustical Society of America* **27**, 247–258 (1955).

[11] H. Kuttruff: *Room Acoustics* (4th ed.). E & FN Spon, London (2000).

[12] D. Lubman: Spatial averaging in a diffuse sound field. *Journal of the Acoustical Society of America* **46**, 532–534 (1969).

[13] R.V. Waterhouse and D. Lubman: Discrete versus continuous space averaging in a reverberant sound field. *Journal of the Acoustical Society of America* **48**, 1–5 (1970).

[14] D. Lubman, R.V. Waterhouse and C. Chien: Effectiveness of continuous spatial averaging in a diffuse sound field. *Journal of the Acoustical Society of America* **53**, 650–659 (1973).

[15] M.R. Schroeder: Frequency-correlation functions of frequency responses in rooms. *Journal of the Acoustical Society of America* **34**, 1819–1823 (1962).

[16] D. Lubman: Fluctuation of sound with position in a reverberant room. *Journal of the Acoustical Society of America* **44**, 1491–1502 (1968).

[17] M.R. Schroeder: Effect of frequency and space averaging on the transmission responses of multimode media. *Journal of the Acoustical Society of America* **46**, 277–283 (1969).

[18] R.L. Weaver and J. Burkhardt: Weak Anderson localization and enhanced backscatter in reverberation rooms and quantum dots. *Journal of the Acoustical Society of America* **96**, 3186–3190 (1994).

[19] F. Jacobsen and A. Rodríguez Molares: The ensemble variance of pure-tone measurements in reverberation rooms. *Journal of the Acoustical Society of America* **127**, 233–237 (2010).

[20] R.H. Lyon: Statistical analysis of power injection and response in structures and rooms. *Journal of the Acoustical Society of America* **45**, 545–565 (1969).

[21] J.L. Davy: The relative variance of the transmission function of a reverberation room. *Journal of Sound and Vibration* **77**, 455–479 (1981).

[22] R.L. Weaver: On the ensemble variance of reverberation room transfer functions, the effect of spectral rigidity. *Journal of Sound and Vibration* **130**, 487–491 (1989).

[23] T.A. Brody, J. Flores, J.B. French, P.A. Mello, A. Pandey and S.S.M. Wong: Random matrix physics: Spectrum and strength fluctuations. *Reviews of Modern Physics* **53**, 385–479 (1981).

[24] F. Jacobsen and H. Ding: Observations on the systematic deviations between two methods of measuring sound transmission loss. *Building Acoustics* **3**, 1–11 (1997).

[25] F. Jacobsen and E. Tiana Roig: Measurement of the sound power incident on the walls of a reverberation room with near field acoustic holography. *Acta Acustica united with Acustica* **96**, 76–91 (2010).

[26] F. Jacobsen: A note on acoustic decay measurements. *Journal of Sound and Vibration*, **115**, 163–170 (1987).

[27] M. Kob and M. Vorländer: Band filters and short reverberation times. *Acustica united with Acta Acustica* **86**, 350–357 (2000).

[28] F.T. Agerkvist and F. Jacobsen: Sound power determination in reverberation rooms at low frequencies. *Journal of Sound and Vibration* **166**, 179–190 (1993).

[29] T.W. Bartel: Effects of absorber geometry on apparent absorption coefficients as measured in a reverberation chamber. *Journal of the Acoustical Society of America* **69**, 1065–1074 (1981).

[30] R.E. Halliwell: Inter-laboratory variability of sound absorption measurements. *Journal of the Acoustical Society of America* **73**, 880–886 (1983).

[31] ISO 140–3: Acoustics – Measurement of sound insulation in buildings and of building elements. Part 3: Laboratory measurements of airborne sound insulation of building elements (1995).

9

Sound Radiation and Scattering

9.1 Introduction

Noise can be generated by many different mechanisms. Vibrating surfaces, combustion, and jet flows are examples of fundamentally different noise generating mechanisms. Insight into the problem is gained by studying elementary sources that are much smaller than the wavelength of the sound they produce; therefore Section 9.2 is devoted to a study of point sources and their interaction. Many practical sound sources involve vibrating surfaces, and in Sections 9.3, 9.4 and 9.5 we shall study radiation of sound by cylindrical, spherical and plane sources. The analytical treatment of scattering of sound by a rigid body is very similar to the analysis of radiation of sound by a body with a prescribed velocity; therefore Sections 9.3.2 and 9.4.2 are devoted to scattering by cylinders and spheres.

9.2 Point Sources

The simplest source to deal with mathematically is a vanishingly small pulsating sphere with a finite volume velocity. Such a source is called a monopole, a point source or a simple source. Any source that changes its volume as a function of time may be approximated by a monopole at frequencies where it is small compared with the wavelength. An enclosed loudspeaker is an example of a source of this category (at low frequencies). Since the particle velocity is diverging at the position of the monopole it follows that it is equivalent to time-varying injection of fluid (in the literature often referred to as mass injection), which leads to the conclusion that the outlet of an engine exhaust system also may be regarded as a monopole. Other examples include sirens, pulsating bubbles and unsteady combustion.

As shown in Section 7.6, under free-field conditions the sound pressure at $\mathbf{r}$ generated by a harmonic point source with the volume velocity Q at $\mathbf{r_0}$ is the solution to the inhomogeneous Helmholtz equation

$$\nabla^2 \hat{p}(\mathbf{r}) + k^2 \hat{p}(\mathbf{r}) = -\mathrm{j}\omega\rho Q\delta(\mathbf{r} - \mathbf{r_0})\mathrm{e}^{\mathrm{j}\omega t}. \tag{9.1}$$

Fundamentals of General Linear Acoustics, First Edition. Finn Jacobsen and Peter Møller Juhl.
© 2013 John Wiley & Sons, Ltd. Published 2013 by John Wiley & Sons, Ltd.

Under free-field conditions the solution to Equation (9.1) is

$$\hat{p}(\mathbf{r}) = \frac{j\omega\rho\, Q e^{j(\omega t - kR)}}{4\pi R},$$

(9.2)

where

$$R = |\mathbf{r} - \mathbf{r_0}|$$

(9.3)

is the distance to the source. This can be seen by inserting Equation (9.2) in Equation (9.1) and integrating over a small spherical volume centred at $\mathbf{r_0}$. This gives, for $\varepsilon \to 0$,

$$\int_0^{2\pi}\int_0^\pi\int_0^\varepsilon (\nabla^2 + k^2)\left(\frac{j\omega\rho\, Q e^{(j\omega t - kR)}}{4\pi R}\right) R^2 \sin\theta\, \mathrm{d}R\mathrm{d}\theta\mathrm{d}\varphi$$

$$= \int_0^{2\pi}\int_0^\pi \frac{\mathrm{d}}{\mathrm{d}R}\left(\frac{j\omega\rho\, Q e^{(j\omega t - kR)}}{4\pi R}\right) R^2 \sin\theta\mathrm{d}\theta\mathrm{d}\varphi\Bigg|_{R=\varepsilon}$$

$$+ 4\pi k^2 \int_0^\varepsilon \frac{j\omega\rho\, Q e^{(j\omega t - kR)}}{4\pi R} R^2\, \mathrm{d}R$$

(9.4)

$$= j\omega\rho Q e^{j\omega t}\left(\varepsilon^2 \frac{\mathrm{d}}{\mathrm{d}R}\left(\frac{e^{-jkR}}{R}\right)\Bigg|_{R=\varepsilon} + k^2 \int_0^\varepsilon Re^{-jkR}\mathrm{d}R\right) \to -j\omega\rho Q e^{j\omega t},$$

where the volume integral of the divergence of the gradient in the second line has been replaced by the corresponding surface integral of the normal component of the gradient (Gauss's theorem). The second volume integral vanishes when $\varepsilon \to 0$. The result is seen to be in agreement with the result of integrating the right-hand side of Equation (9.1) over the small volume.

Example 9.1 As we have seen in Section 7.6 the *Green's function* is the solution of the inhomogeneous Helmholtz equation,

$$(\nabla^2 + k^2)G(\mathbf{r}, \mathbf{r_0}) = -\delta(\mathbf{r} - \mathbf{r_0}).$$

This is the sound pressure at $\mathbf{r}$ generated by a point source at $\mathbf{r_0}$, normalised by letting $j\omega\rho Q = 1$. This corresponds to a point source with a frequency independent *volume acceleration*, as mentioned in Section 7.6. Note that $G(\mathbf{r}, \mathbf{r_0}) = G(\mathbf{r_0}, \mathbf{r})$ irrespective of the boundary conditions, in agreement with the reciprocity principle. The free-space Green's function follows from Equation (9.2)[1]:

$$G(\mathbf{r}, \mathbf{r_0}) = \frac{e^{-jkR}}{4\pi R}.$$

---

[1] Some authors have a factor of 4π on the right-hand side of the inhomogeneous Helmholtz equation that defines the Green's function, which leads to a free-field Green's function of the form $G(\mathbf{r}, \mathbf{r_0}) = e^{-jkR}/R$. It should also be remembered that some authors use the $e^{-i\omega t}$ sign convention, which changes the sign of the phase.

Example 9.2 It is interesting to compare the free-space Greens function with the earlier derived Greens functions in ducts and rooms at low frequencies. The Greens function for a semi-infinite lossless duct is given by Equation (7.166),

$$G(\mathbf{r}, \mathbf{r}_0) = -\frac{j}{S} \sum_m \frac{\psi_m(\mathbf{r})\psi_m(\mathbf{r}_0)}{\sqrt{k^2 - k_m^2}} e^{-jk_{zm}z},$$

and the Greens function for a rectangular lossless room is given by Equation (8.34),

$$G(\mathbf{r}, \mathbf{r}_0) = -\frac{1}{V} \sum_m \frac{\psi_m(\mathbf{r})\psi_m(\mathbf{r}_0)}{k^2 - k_m^2}.$$

It is apparent that a monopole with the volume velocity $Qe^{j\omega t}$ at very low frequencies will generate the sound pressure

$$\hat{p}(r) = \begin{cases} \frac{j\omega\rho Q}{4\pi R} e^{j(\omega t - kR)} & \text{in free space,} \\ \frac{\rho c Q}{S} e^{j(\omega t - kz)} & \text{in a semi-infinite duct,} \\ \frac{\rho c^2 Q}{j\omega V} e^{j\omega t} & \text{in a room.} \end{cases}$$

Note that the radiation impedance in free space is mass-like, that the radiation impedance in the duct is real-valued at low frequencies where only plane waves can propagate, and that the air in an enclosure at low frequencies behaves like an elastic spring.

If the monopole is situated at the origin of the coordinate system Equation (9.2) becomes

$$\hat{p}(r) = \frac{j\omega\rho Q e^{j(\omega t - kr)}}{4\pi r}. \tag{9.5}$$

The radial component of the particle velocity can be calculated from Euler's equation of motion,

$$\hat{u}_r(r) = -\frac{1}{j\omega\rho} \frac{\partial \hat{p}(r)}{\partial r} = \frac{jkQ e^{j(\omega t - kr)}}{4\pi r}\left(1 + \frac{1}{jkr}\right) = \frac{\hat{p}(r)}{\rho c}\left(1 + \frac{1}{jkr}\right). \tag{9.6}$$

Note that

$$4\pi r^2 \hat{u}_r(r) = jkrQ e^{j(\omega t - kr)}\left(1 + \frac{1}{jkr}\right) \rightarrow Qe^{j\omega t} \tag{9.7}$$

in the limit of $kr \rightarrow 0$, just as we would expect.

The sound intensity generated by the monopole can be determined from Equations (9.5) and (9.6):

$$I_r(r) = \frac{1}{2}\text{Re}\{\hat{p}(r)\hat{u}_r^*(r)\} = \frac{1}{2}\text{Re}\left\{\frac{j\rho\omega Q}{4\pi r}\frac{-jkQ^*}{4\pi r}\left(1 - \frac{1}{jkr}\right)\right\} = \frac{\rho(\omega|Q|)^2}{32\pi^2 r^2 c}. \tag{9.8}$$

The sound power radiated by the monopole can be obtained by integrating the sound intensity over a spherical surface. Because of the spherical symmetry this integration corresponds to multiplying with the area of the surface,

$$P_a = \frac{\rho(\omega|Q|)^2}{32\pi^2 cr^2} 4\pi r^2 = \frac{\rho c k^2 |Q|^2}{8\pi}. \tag{9.9}$$

Note that the sound power is proportional to the square of the frequency, indicating poor radiation at low frequencies.

An alternative way of deriving the sound power places the integration surface infinitely close to the source. The surface integral of the particle velocity is the volume velocity (cf. Equation (9.7)), and therefore,

$$
\begin{aligned}
P_a &= \frac{1}{2} \lim_{r \to 0} \text{Re}\{\hat{p}(r) Q^* e^{-j\omega t}\} = \frac{1}{2} \lim_{r \to 0} \text{Re}\left\{ \frac{j\omega\rho|Q|^2 e^{-jkr}}{4\pi r} \right\} \\
&= \frac{\rho c k^2 |Q|^2}{8\pi} \lim_{r \to 0} \text{Re}\left\{ \frac{j\cos(kr) + \sin(kr)}{kr} \right\} = \frac{\rho c k^2 |Q|^2}{8\pi}.
\end{aligned}
\tag{9.10}
$$

This method of determining the sound power may seem to be more complicated, but it has the advantage of being more general, since it allows the sound pressure to include contributions from reflecting surfaces or from other sources. We shall make use of this method in what follows.

It is a simple matter to calculate the sound field generated by a combination of point sources. Since they are infinitely small they do not disturb each other, and therefore the sound fields are simply added.

The combination of two identical monopoles placed close to each other and expanding and contracting in antiphase is a point dipole; see Figure 9.1 The sound pressure generated by the two monopoles in free space is[2]

$$\hat{p}(r, \theta) = \frac{j\rho\omega Q e^{j(\omega t - kr_1)}}{4\pi r_1} - \frac{j\rho\omega Q e^{j(\omega t - kr_2)}}{4\pi r_2}. \tag{9.11}$$

Any small oscillating body with a fixed volume acts as a point dipole. An unenclosed loudspeaker, for example, is a dipole at low frequencies.

If $2h \ll \lambda$ we can replace the finite difference with the gradient multiplied by $2kh$,

$$\hat{p}(r, \theta) = 2kh \cos\theta \frac{\partial}{\partial r}\left(\frac{j\rho\omega Q e^{j(\omega t - kr)}}{4\pi r} \right) = \frac{\rho c h k^2 Q}{2\pi r}\left(1 + \frac{1}{jkr} \right) \cos\theta e^{j(\omega t - kr)}. \tag{9.12}$$

The factor in parenthesis is a near field term (cf. Equations (3.40) and (9.6)). Note that the sound pressure is proportional to $h|Q|$, varies as $\cos\theta$ and is identically zero in the plane between the two monopoles.

---

[2] It is apparent that the dipole produces no net volume outflow. On the other hand it exerts a *force* on the medium. It can be shown that the force is $\mathbf{F} = j\omega\rho Q (\mathbf{r}_1 - \mathbf{r}_2)$ if the distance between the two monopoles is much shorter than the wavelength. Accordingly, Euler's equation of motion becomes

$$\nabla \hat{p} + j\omega\rho \mathbf{u} = \mathbf{F} e^{j\omega t} \delta(\mathbf{r} - \mathbf{r}_0),$$

from which it follows that the corresponding inhomogenous wave equation has the form

$$(\nabla^2 + k^2)\hat{p} = \nabla \cdot (\mathbf{F} e^{j\omega t} \delta(\mathbf{r} - \mathbf{r}_0)).$$

Here $\mathbf{r}_0$ is the position midway between the two monopoles [1].

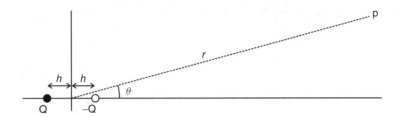

Figure 9.1   Determining the far field of a point dipole

The sound power of the dipole can be calculated by integrating the mean square sound pressure over a spherical surface centred midway between the two monopoles:

$$
\begin{aligned}
P_{\mathrm{a}} &= \int_0^\pi \int_0^{2\pi} \frac{|\hat{p}(r,\theta)|^2}{2\rho c} r^2 \sin\theta \, \mathrm{d}\varphi \mathrm{d}\theta = \frac{\rho c h^2 k^4 |Q|^2}{4\pi} \int_0^\pi \cos^2\theta \sin\theta \, \mathrm{d}\theta \\
&= \frac{\rho c h^2 k^4 |Q|^2}{4\pi} \int_{-1}^1 x^2 \mathrm{d}x = \frac{\rho c h^2 k^4 |Q|^2}{6\pi}.
\end{aligned}
\tag{9.13}
$$

Alternatively we can make use of the method we used in deriving Equation (9.10),

$$
\begin{aligned}
P_{\mathrm{a}} &= 2 \cdot \frac{1}{2} \lim_{\substack{r_1 \to 0 \\ r_2 \to 2h}} \mathrm{Re}\left\{ \left(\frac{j\omega\rho \, Q \mathrm{e}^{-jkr_1}}{4\pi r_1} - \frac{j\omega\rho \, Q \mathrm{e}^{-jkr_2}}{4\pi r_2} \right) Q^* \right\} \\
&= \frac{\rho c k^2 |Q|^2}{4\pi} \left(1 - \frac{\sin 2kh}{2kh} \right) \simeq \frac{\rho c k^2 |Q|^2}{4\pi} \frac{(2kh)^2}{6} = \frac{\rho c h^2 k^4 |Q|^2}{6\pi}.
\end{aligned}
\tag{9.14}
$$

Note that the sound power of the dipole is proportional to the fourth power of the frequency, indicating very poor sound radiation at low frequencies. The physical explanation of the poor radiation efficiency of the dipole is that the two monopoles almost cancel each other, the more so the closer they are in terms of the wavelength.

Example 9.3 The sound power of a monopole is affected by the presence of a plane, rigid surface, which acts as an image source. Assuming that the distance to the plane is h we can calculate the resulting sound power in the same way as we used in deriving Equation (9.14):

$$
\begin{aligned}
P_{\mathrm{a}} &= \frac{1}{2} \lim_{\substack{r_1 \to 0 \\ r_2 \to 2h}} \mathrm{Re}\left\{ \left(\frac{j\omega\rho \, Q \mathrm{e}^{-jkr_1}}{4\pi r_1} + \frac{j\omega\rho \, Q \mathrm{e}^{-jkr_2}}{4\pi r_2} \right) Q^* \right\} \\
&= \frac{\rho c k^2 |Q|^2}{8\pi} \left(1 + \frac{\sin 2kh}{2kh} \right).
\end{aligned}
$$

The factor in parentheses is shown in Figure 9.2. It is apparent that the sound power is doubled if $kh \ll 1$. However, the rigid surface is seen to have an insignificant influence on the sound power output of the source when h exceeds a quarter of a wavelength.

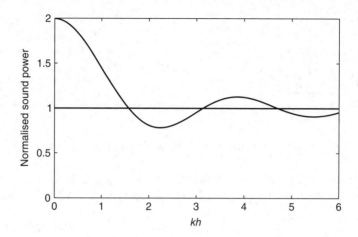

Figure 9.2 The influence of an infinite, plane, rigid surface on the sound power output of a monopole

Two dipoles of equal but opposite strength make a quadrupole. Two examples of such combinations are shown in Figure 9.3. The importance of quadrupoles is mainly due to the fact that the momentum flux in airflows, which is the main source mechanism in unsteady airflow in the absence of solid bodies, is a quadrupole.

The sound pressure generated by a longitudinal quadrupole at the origin of the coordinate system is

$$\hat{p}(r,\theta) = -\frac{j\rho c h^2 k^3 Q}{\pi r} e^{j(\omega t - kr)} \left(\cos^2\theta \left(1 + \frac{3}{jkr} - \frac{3}{(kr)^2} \right) - \frac{1}{jkr} + \frac{1}{(kr)^2} \right), \quad (9.15)$$

where $Q(2h)^2$ is the strength, as easily shown by superposition. The sound power is

$$P_a = \frac{2\rho c h^4 k^6 |Q|^2}{5\pi}. \quad (9.16)$$

Note that the sound power is proportional to the sixth power of the frequency. Accordingly, quadrupoles are very inefficient radiators of sound at low frequencies. A tuning fork is, to a good approximation, a longitudinal quadrupole. The sound produced by a tuning fork is not radiated by the vibrating prongs but by the stem moving (much less) up and down, thereby exciting a solid surface.

Figure 9.3 A longitudinal and a lateral quadrupole

A lateral quadrupole at the origin generates the sound pressure

$$\hat{p}(r, \theta, \phi) = -\frac{j\rho c h^2 k^3 Q}{\pi r} e^{j(\omega t - kr)} \cos\theta \sin\theta \cos\phi \left(1 + \frac{3}{jkr} - \frac{3}{(kr)^2}\right). \tag{9.17}$$

The corresponding sound power is

$$P_a = \frac{2\rho c h^4 k^6 |Q|^2}{15\pi} \tag{9.18}$$

(see Example 9.5 in Section 9.2.2).

9.2.1 Reciprocity

A monopole with the volume velocity Q at $\mathbf{r_0}$ generates the sound pressure $\hat{p}(\mathbf{r})$ and the particle velocity $\hat{\mathbf{u}}(\mathbf{r})$. If the monopole is moved to another position, $\mathbf{r_0'}$, the sound pressure and the particle velocity become $\hat{p}'(\mathbf{r})$ and $\hat{\mathbf{u}}'(\mathbf{r})$. Let us examine the divergence of $\hat{p}\hat{\mathbf{u}}' - \hat{p}'\hat{\mathbf{u}}$,

$$\begin{aligned}
\nabla \cdot (\hat{p}\hat{\mathbf{u}}' - \hat{p}'\hat{\mathbf{u}}) &= \nabla\hat{p} \cdot \hat{\mathbf{u}}' + \hat{p}\nabla \cdot \hat{\mathbf{u}}' - \nabla\hat{p}' \cdot \hat{\mathbf{u}} - \hat{p}'\nabla \cdot \hat{\mathbf{u}} \\
&= -j\omega\rho\hat{\mathbf{u}} \cdot \hat{\mathbf{u}}' - \frac{j\omega}{\rho c^2}\hat{p}\hat{p}' + \hat{p}Qe^{j\omega t}\delta(\mathbf{r} - \mathbf{r_0'}) \\
&\quad + j\omega\rho\hat{\mathbf{u}}' \cdot \hat{\mathbf{u}} + \frac{j\omega}{\rho c^2}\hat{p}'\hat{p} - \hat{p}'Qe^{j\omega t}\delta(\mathbf{r} - \mathbf{r_0}) \\
&= Qe^{j\omega t}(\hat{p}\delta(\mathbf{r} - \mathbf{r_0'}) - \hat{p}'\delta(\mathbf{r} - \mathbf{r_0})),
\end{aligned} \tag{9.19}$$

where use has been made of Equation (2.12) and Euler's equation of motion,

$$\nabla\hat{p} + j\omega\rho\hat{\mathbf{u}} = 0. \tag{9.20}$$

Let us integrate this equation over the volume of interest. The result is

$$\int_V \nabla \cdot (\hat{p}\hat{\mathbf{u}}' - \hat{p}'\hat{\mathbf{u}})dV = Qe^{j\omega t}(\hat{p}(\mathbf{r_0'}) - \hat{p}'(\mathbf{r_0})). \tag{9.21}$$

In an interior problem the region of interest is the volume of the enclosure; in an exterior problem we will integrate over the volume enclosed by a spherical (or hemispherical) surface in the far field where it may be assumed that

$$\hat{u}_r(\mathbf{r}) = \frac{\hat{p}(\mathbf{r})}{\rho c}. \tag{9.22}$$

If the walls of the enclosure or the boundaries of an outdoor propagation problem can be assumed to be locally reacting the boundary condition is

$$\frac{\hat{p}}{\hat{u}_n} = Z. \tag{9.23}$$

If use is made of Gauss's theorem the integral on the left-hand side of Equation (9.21) becomes

$$\int_S (\hat{p}\hat{\mathbf{u}}' - \hat{p}'\hat{\mathbf{u}}) \cdot \mathbf{n}\, dS = \int_S \left(\frac{\hat{p}\hat{p}'}{Z} - \frac{\hat{p}'\hat{p}}{Z}\right)dS = 0. \tag{9.24}$$

All in all it follows that

$$\hat{p}(\mathbf{r}_0') = \hat{p}'(\mathbf{r}_0), \tag{9.25}$$

which shows that the sound pressure at position $\mathbf{r}_0'$ generated by a monopole at $\mathbf{r}_0$ is identical with the sound pressure at $\mathbf{r}_0$ generated by the monopole when it is moved to $\mathbf{r}_0'$. In other words, when source and listener positions are interchanged the sound pressure is the same, and the Green's function is symmetrical in $\mathbf{r}$ and $\mathbf{r}_0$,

$$G(\mathbf{r}, \mathbf{r}_0) = G(\mathbf{r}_0, \mathbf{r}). \tag{9.26}$$

This is known as the reciprocity principle, and it holds true also in a region defined by non-locally reacting surfaces such as membranes, shells and plates [1].

9.2.2 Sound Power Interaction of Coherent Sources

Uncorrelated sources do not affect the sound power emission of each other. However, as demonstrated by Example 9.3 correlated (or coherent) sources *do* affect each other. In this section we shall study how the sound power output of a monopole is affected by the presence of another monopole that emits sound at the same frequency. The two sources have volume velocities $Q_1 e^{j\omega t}$ and $Q_2 e^{j\omega t}$ and are separated by a distance of R. With the same notation as used in the derivation of Equation (9.14) the sound power emitted by the first monopole is

$$
\begin{aligned}
P_{a1} &= \frac{1}{2} \lim_{\substack{r_1 \to 0 \\ r_2 \to R}} \mathrm{Re}\left\{ \left(\frac{j\omega\rho Q_1 e^{-jkr_1}}{4\pi r_1} + \frac{j\omega\rho Q_2 e^{-jkr_2}}{4\pi r_2} \right) Q_1^* \right\} \\
&= \frac{\rho c k^2 |Q_1|^2}{8\pi} \left(1 + \mathrm{Re}\left\{ \frac{Q_2}{Q_1} \frac{j e^{-jkR}}{kR} \right\} \right) \\
&= \frac{\rho c k^2 |Q_1|^2}{8\pi} \left(1 + \frac{|Q_2|}{|Q_1|} \frac{\sin(kR + \varphi_{12})}{kR} \right),
\end{aligned}
\tag{9.27}
$$

where φ_{12} is the phase angle of Q_1/Q_2. This can take any value, from which it follows that the sound power can be increased or decreased by the presence of the second source. In fact the sound power can be negative, so that the source becomes a *sink* (a source that absorbs sound power).

The sound power of the second source is

$$
\begin{aligned}
P_{a2} &= \frac{1}{2} \lim_{\substack{r_1 \to R \\ r_2 \to 0}} \mathrm{Re}\left\{ \left(\frac{j\omega\rho Q_1 e^{-jkr_1}}{4\pi r_1} + \frac{j\omega\rho Q_2 e^{-jkr_2}}{4\pi r_2} \right) Q_2^* \right\} \\
&= \frac{\rho c k^2 |Q_2|^2}{8\pi} \left(1 + \mathrm{Re}\left\{ \frac{Q_1}{Q_2} \frac{j e^{-jkR}}{kR} \right\} \right) \\
&= \frac{\rho c k^2 |Q_2|^2}{8\pi} \left(1 + \frac{|Q_1|}{|Q_2|} \frac{\sin(kR - \varphi_{12})}{kR} \right),
\end{aligned}
\tag{9.28}
$$

and the total sound power is

$$P_a = \frac{\rho c k^2}{8\pi} \left(|Q_1|^2 + |Q_2|^2 + |Q_1||Q_2| \frac{\sin(kR + \varphi_{12}) + \sin(kR - \varphi_{12})}{kR} \right)$$

$$= \frac{\rho c k^2}{8\pi} \left(|Q_1|^2 + |Q_2|^2 + 2|Q_1||Q_2| \frac{\sin kR}{kR} \cos \varphi_{12} \right). \tag{9.29}$$

Example 9.4 With $|Q_1| = |Q_2| = |Q|$ and $\varphi_{12} = \pi$ Equation (9.29) becomes

$$P_a = \frac{\rho c k^2 |Q|^2}{4\pi} \left(1 - \frac{\sin kR}{kR} \right),$$

which is seen to agree with Equation (9.14) as of course it should. In the limit of $kR \ll 1$ we have a dipole.

Active noise control involves generating a secondary sound field that interferes destructively with the sound field produced by the primary source so that the noise is reduced. The secondary sound field is generated using a controlled source (or several controlled sources).

If we assume that monopole no 1 is the primary, given source, whereas source no 2 can be controlled, then possible control strategies could be i) to drive the second source in such a way that the resulting sound pressure at a certain position is zero, ii) to maximise the sound power absorbed by the second source, and iii) to minimise the total radiated sound power.

It will always be possible to obtain a very low sound pressure at a given position by adjusting the amplitude and phase of the secondary source appropriately. However, this method will in general tend to increase the sound pressure at other positions than the selected one, so its usefulness is limited. Examples of the resulting directional pattern are presented in [2].

The second method is more interesting. It is clear from Equation (9.28) that the phase should be

$$\varphi_{12} = \frac{\pi}{2} + kR. \tag{9.30}$$

The minimum is then found by differentiating the resulting equation with respect to $|Q_2|$. The result is

$$|Q_2| = \frac{|Q_1|}{2kR}. \tag{9.31}$$

Inserting into Equation (9.28) gives

$$P_{a2} = \frac{\rho c k^2 |Q_1|^2}{8\pi} \left(\frac{1}{4(kR)^2} - \frac{1}{2(kR)^2} \right) = -\frac{\rho c k^2 |Q_1|^2}{32\pi(kR)^2}, \tag{9.32}$$

which shows that the secondary source can absorb a substantial amount of sound power from the primary source at low frequencies – much more than its free-field power, in fact.

However, inserting Equation (9.31) into Equations (9.27) and (9.29) gives

$$P_{a1} = \frac{\rho c k^2 |Q_1|^2}{8\pi} \left(1 + \frac{\cos 2kR}{2(kR)^2} \right) = \frac{\rho c k^2 |Q_1|^2}{8\pi} \left(1 - \left(\frac{\sin kR}{kR} \right)^2 + \frac{1}{2(kR)^2} \right), \quad (9.33)$$

$$P_a = \frac{\rho c k^2 |Q_1|^2}{8\pi} \left(1 - \left(\frac{\sin kR}{kR} \right)^2 + \frac{1}{4(kR)^2} \right). \quad (9.34)$$

Examination of these equations leads to the conclusion that the secondary source at low frequencies will increase not only the sound power output of the primary source, but also the total sound power, from which it follows that maximising the power absorbed by the secondary source is a poor noise control strategy.

It is easy to show that the total radiated sound power is minimised if

$$\varphi_{12} = \pi, \quad (9.35)$$

and

$$|Q_2| = |Q_1| \frac{\sin kR}{kR}. \quad (9.36)$$

Inserting into Equation (9.29) gives

$$P_a = \frac{\rho c k^2 |Q_1|^2}{8\pi} \left(1 - \left(\frac{\sin kR}{kR} \right)^2 \right). \quad (9.37)$$

At low frequencies this expression is very close to the expression for the sound power of a dipole (cf. Equation (9.14)).

Figure 9.4 shows the ratio of the free-field power of the primary source to the minimised total sound power. It is apparent that the distance between the primary and the secondary

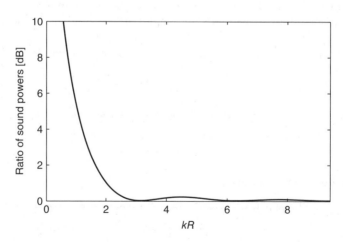

Figure 9.4 Ratio of the free-field power of the primary source to the minimum value of the total sound power of the primary and secondary source

source must be less than a quarter of a wavelength for any substantial reduction to occur, which leads to the conclusion that the potential of active control under free-field conditions is quite limited.

Inserting Equations (9.35) and (9.36) into Equation (9.28) gives an interesting result: $P_{a2} = 0$. The secondary source does not emit any sound power at all, nor does it absorb any sound power. Its effect on the sound power output of the primary source is due to impedance mismatch.

Example 9.5 The sound power of the longitudinal and the lateral quadrupoles described in the foregoing (Equations (9.16) and (9.18)) can with advantage be calculated using Equation (9.28). For example, each of the four monopoles composing the lateral quadrupole is exposed to the sound pressure from another monopole in phase and two monopoles in antiphase. All in all the quadrupole emits the sound power

$$P_a = \frac{4\rho c k^2 |Q|^2}{8\pi} \lim_{kh \to 0} \left(1 - 2\frac{\sin(2kh)}{2kh} + \frac{\sin(2\sqrt{2}kh)}{2\sqrt{2}kh} \right)$$

$$= \frac{2\rho c h^4 k^6 |Q|^2}{15\pi}.$$

9.2.3 Fundamentals of Beamforming

The purpose of beamforming is to improve the directivity and thus increase the signal-to-noise ratio compared with that of one single acoustic transducer by combining several transducers (usually pressure microphones) in an array. Possible applications include video conferences (focus on the speaker), hearing aids (suppress disturbing background noise), hands-free telephones (suppress background noise), environmental surveillance, etc. In its most fundamental version it is a method of mapping incoming plane waves coming from various directions with an array of pressure microphones, the directivity of which is steered in different directions by aligning the signals in time. This is known as delay and sum beamforming [3]. It is easy to show that a beamformer composed by a line array with M equidistantly arranged pressure microphones ($m = 1, 2, 3, \ldots, M$) (a uniform line array) will be focused in the direction θ (the angle to the normal of the array) with delays

$$\tau_m = \frac{(m-1)d \sin \theta}{c}, \tag{9.38}$$

where d is the spacing between adjacent microphones; see Figure 9.5. It is also easy to show that the output of such a beamformer produced by a plane wave with amplitude A incident at the angle θ_1 is

$$y(\theta_1) = A e^{j\omega t} \sum_{m=1}^{M} e^{-j\omega(\tau_m - (m-1)d \sin \theta_1 / c)}. \tag{9.39}$$

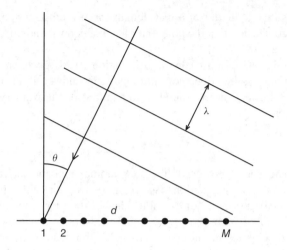

Figure 9.5   A uniform line array exposed to a plane wave

Inspection of Equation (9.39) shows that the output of the beamformer (the beampattern) assumes its maximum when a plane wave arrives from the direction at which it is focused ($\theta_1 = 30°$). (A similar steered radiation directivity can be obtained, at least at low frequencies, by driving an array of monopoles (approximated by loudspeakers) with appropriate individual delays.)

The 3-dB width of the 'main lobe' defines the resolution. However, as can be seen in Figure 9.6, there are a number of secondary, smaller maxima (usually referred to as 'side lobes'). Therefore, it is convenient to evaluate the beamformer response not only by its resolution but also by means of the difference in level between the peak of the highest side lobe and the peak of the main lobe. Moreover, note that aliasing (due to spatial

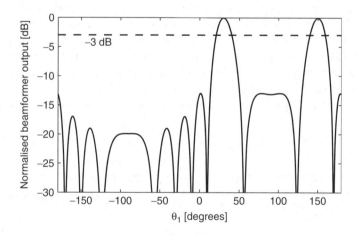

Figure 9.6   Beampattern of a uniform line array exposed to a plane wave. Focus at 30° (and 150°); $M = 10$; $d = 0.3\lambda$

undersampling) will occur (see Section B.7) when $d > \lambda/2$, resulting in 'false' responses from directions without sources.

The performance of a beamformer depends strongly on the array geometry (shape, size, number of microphones, microphone distribution). For example, planar arrays obviously have the same response to sound coming from the two sides, and planar arrays with a uniform distribution of microphones have much finer spatial resolution in the perpendicular direction than in the direction of the array. Moreover, the microphones can be arranged equidistantly, logarithmically or in a pseudo-random manner, and they can be suspended in free space [4] or mounted on solid bodies [5]. Arrays with the microphones uniformly mounted in a circular arrangement have a similar directivity in all directions in the plane of the array [4], whereas arrays with the microphones mounted on a spherical surface can achieve a similar directivity in all directions [6].

Beamformers can also be focused at positions near the array, and thus the focus can be scanned over a large three-dimensional region, but the method cannot *reconstruct* sound fields (unlike acoustic holography; see Section 9.5.3). A fundamental assumption of beamforming is that the sources that generate the sound field are uncorrelated.

Over the years numerous more refined methods than the simple delay-and-sum have been developed. The simplest is to modify the delay-and-sum by applying different weights on the microphone signals so as to modify the directivity [3]. Eigenbeamforming is based on an expansion in circular harmonics (for circular arrays in free space, mounted on a cylinder or mounted on a sphere) [7] and spherical harmonics (for spherical arrays) [8]. However, the delay-and-sum method is still widely used because it is more robust than the more refined methods. To this can be added that various signal processing methods for improving the resolution or for suppressing side lobes can be applied. One example of such signal processing is spatial deconvolution, which, however, is extremely time-consuming unless spectral methods can be applied, which requires that the response to a monopole, the point spread function, is shift invariant, i.e., only dependent on the distance between the current observation point and the source position and not on the individual positions [9].

9.3 Cylindrical Waves

9.3.1 Radiation from Cylindrical Sources

Some sources of sound such as e.g. the vibrating string of a violin can be approximated by a (circular) cylindrical source of infinite extent, which facilates an analytical solution. By contrast, the problem of radiation from finite cylinders cannot be solved analytically in general. Other cylindrical sources include wind turbine towers and ventilation systems. Hence, in addition to being among the most simple radiation problems for which an analytical solution exists, cylindrical sources are also of some practical interest, even though real sources are not of infinite extent, of course.

In cylindrical coordinates as defined in Figure 9.7 the Helmholtz equation becomes (cf. Equation (2.20b))

$$\frac{\partial^2 \hat{p}}{\partial r^2} + \frac{1}{r}\frac{\partial \hat{p}}{\partial r} + \frac{1}{r^2}\frac{\partial^2 \hat{p}}{\partial \varphi^2} + \frac{\partial^2 \hat{p}}{\partial z^2} + k^2 \hat{p} = 0. \tag{9.40}$$

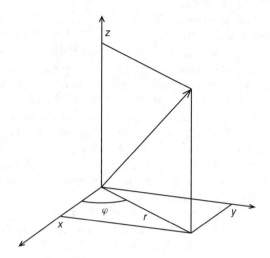

Figure 9.7 Cylindrical coordinate system

This equation can be solved using separation of the variables. It is assumed that the sound pressure may be written as the product of functions that each depends only on one variable:

$$\hat{p}(r, \varphi, z, t) = p_r(r) p_\varphi(\varphi) p_z(z) e^{j\omega t}. \tag{9.41}$$

If the time dependence is omitted, inserting Equation (9.41) into Equation (9.40), and dividing by $p_r(r) p_\varphi(\varphi) p_z(z)$ leads to

$$\left(\frac{1}{p_r} \frac{\partial^2 p_r}{\partial r^2} + \frac{1}{p_r r} \frac{\partial p_r}{\partial r} + \frac{1}{p_\varphi r^2} \frac{\partial^2 p_\varphi}{\partial \varphi^2} \right) + \left(\frac{1}{p_z} \frac{\partial^2 p_z}{\partial z^2} \right) = -k^2. \tag{9.42}$$

Since the terms in the first bracket depend on r and φ only and the terms in the second bracket depend on z only, each of the brackets must equal a constant, $-k_r^2$ and $-k_z^2$ respectively, so that $k^2 = k_r^2 + k_z^2$. Furthermore, multiplying the first bracket with r^2 gives

$$\frac{r^2}{p_r} \frac{\partial^2 p_r}{\partial r^2} + \frac{r}{p_r} \frac{\partial p_r}{\partial r} + \frac{1}{p_\varphi} \frac{\partial^2 p_\varphi}{\partial \varphi^2} = -k_r^2 r^2, \tag{9.43}$$

which now leads to the conclusion that the last term on the left-hand side of Equation (9.43) must equal a constant, $-k_\varphi^2$,

$$\frac{1}{p_\varphi} \frac{\partial^2 p_\varphi}{\partial \varphi^2} = -k_\varphi^2, \tag{9.44}$$

or

$$\frac{d^2 p_\varphi}{d\varphi^2} + k_\varphi^2 p_\varphi = 0, \tag{9.45}$$

where the partial derivatives have been replaced with ordinary derivatives. Hence, for p_φ and p_z two regular second order ordinary differential equations arise, for which the general solutions are

$$p_z(z) = Ae^{-jk_z z} + Be^{jk_z z} \tag{9.46}$$

and

$$p_\varphi(\varphi) = Ce^{-jk_\varphi \varphi} + De^{jk_\varphi \varphi} \tag{9.47}$$

respectively.

In the rest of this section it will be assumed that the vibration of the cylindrical surface is independent of z, i.e., that $A = 1$ and that $B = k_z = 0$. This assumption in effect reduces the present problem to two dimensions.

Since the sound field must be periodic in φ, $p_\varphi(\varphi) = p_\varphi(\varphi + 2\pi)$, it can be concluded that $k_\varphi = m$, where m is an integer. Hence Equation (9.43) becomes

$$\frac{d^2 p_r}{dr^2} + \frac{1}{r}\frac{dp_r}{dr} + p_r\left(k_r^2 - \frac{m^2}{r^2}\right) = 0. \tag{9.48}$$

The solutions to this equation are the cylindrical Bessel functions; see Appendix C.1. For the present discussion solutions in terms of the Bessel functions of the third kind (often denoted cylindrical Hankel functions) turns out to be most convenient,

$$p_r(r) = EH_m^{(1)}(k_r r) + FH_m^{(2)}(k_r r), \tag{9.49}$$

since the Hankel functions may be interpreted as outgoing and incoming waves.

The cylindrical Hankel functions are related to the cylindrical Bessel functions of the first and second kind (the latter are also known as the cylindrical Neumann functions; see Figure 9.8) as follows:

$$H_m^{(1)}(k_r r) = J_m(k_r r) + jN_m(k_r r), \tag{9.50}$$

$$H_m^{(2)}(k_r r) = J_m(k_r r) - jN_m(k_r r). \tag{9.51}$$

Consider the asymptotic behaviour of the Hankel functions (see Equations (C.8) and (C.9)):

$$\lim_{r \to \infty} H_m^{(1)}(k_r r) = \sqrt{\frac{2}{\pi k_r r}} e^{j(k_r r - m\pi/2 - \pi/4)} \tag{9.52}$$

and

$$\lim_{r \to \infty} H_m^{(2)}(k_r r) = \sqrt{\frac{2}{\pi k_r r}} e^{-j(k_r r - m\pi/2 - \pi/4)}. \tag{9.53}$$

With the time convention $e^{j\omega t}$ it is apparent that Equation (9.52) represents an incoming wave whereas Equation (9.53) represents an outgoing wave. For a cylinder in free (two-dimensional) space the Sommerfeld radiation condition [1] must be fulfilled, which in effect means that only the outgoing wave can exist. Therefore, it is concluded that $E = 0$, and the total solution to Equation (9.40) with the above-mentioned assumptions and boundary conditions becomes,

$$\hat{p}(r, \varphi) = \sum_{m=-\infty}^{\infty} A_m H_m^{(2)}(k_r r) e^{jm\varphi} e^{j\omega t}. \tag{9.54}$$

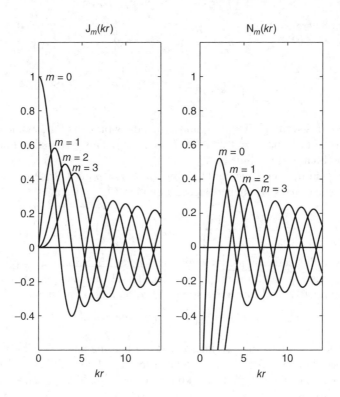

Figure 9.8   Cylindrical Bessel and Neumann functions for $m = 0, 1, 2$ and 3

Example 9.6 The cylindrical Hankel functions diverge in the limit of vanishing arguments. However, since r denotes the distance to the centre of the cylindrical sound source, this limit is never reached for a cylinder of finite radius.

Example 9.7 The description of the sound field inside a duct requires cylindrical Hankel functions of both kinds. Instead Bessel functions of the first and second kind are often used as the two independent solutions of Equation (9.48),

$$p_r(r) = E J_m(k_r r) + F N_m(k_r r).$$

The Bessel function of the second kind, N_m, is as mentioned also called the cylindrical Neumann function, and in some texts denoted by Y_m. The Neumann function diverges in the limit of a vanishing argument and as a consequence $F = 0$ for ducts without a core. The remaining Bessel function of the first kind J_m represents a standing wave, and this term may be interpreted as the interference between an incoming and an outgoing cylindrical wave.

Radiation from a cylindrical source of zeroth order corresponds to examining Equation (9.54) with $m = 0$,

$$\hat{p}(r, \varphi) = \hat{p}(r) = A_0 H_0^{(2)}(kr)e^{j\omega t}, \tag{9.55}$$

where k_r has been replaced by k since $k_z = 0$.

It is apparent that this source is omnidirectional, i.e., that the sound field does not depend on the angular direction, φ. In cylindrical coordinates the gradient is

$$\nabla = \left(\frac{\partial}{\partial r}, \frac{1}{r} \frac{\partial}{\partial \varphi}, \frac{\partial}{\partial z} \right), \tag{9.56}$$

and thus the particle velocity in the radial direction becomes

$$\hat{u}_r(r) = \frac{-1}{j\omega \rho} \frac{\partial \hat{p}}{\partial r} = \frac{-jA_0}{\rho c} H_1^{(2)}(kr)e^{j\omega t}, \tag{9.57}$$

(see Appendix C.1 for calculation of the derivatives of the Hankel functions). Hence, for this source the radial velocity is independent of the angular coordinate, which means that this cylindrical source is an infinitely long pulsating cylinder. If the radial velocity of the cylinder at the surface $r = a$ is $U_0 e^{j\omega t}$ a relation between U_0 and A_0 is easily obtained by evaluating Equation (9.57) at $r = a$,

$$A_0 = U_0 \frac{j\rho c}{H_1^{(2)}(ka)}. \tag{9.58}$$

Example 9.8 In the far field the sound field locally resembles a plane wave, since the ratio of Equation (9.55) to Equation (9.57),

$$\frac{\hat{p}}{\hat{u}_r} = j\rho c \frac{H_0^{(2)}(kr)}{H_1^{(2)}(kr)} \to \rho c \text{ for } kr \to \infty,$$

cf. Equation (9.53).

The radial sound intensity is

$$I_r = \frac{1}{2}\text{Re}\{\hat{p}\hat{u}_r^*\} = \frac{1}{2}\frac{|A_0|^2}{\rho c}\text{Re}\{jH_0^{(2)}(kr)H_1^{(1)}(kr)\}$$
$$= \frac{-1}{2}\frac{|A_0|^2}{\rho c}\text{Im}\{H_0^{(2)}(kr)H_1^{(1)}(kr)\}, \tag{9.59}$$

since the complex conjugate of $H_m^{(2)}$ is $H_m^{(1)}$ and *vice versa*. In the far field Equation (9.59) becomes

$$\lim_{kr \to \infty} I_r = \frac{|A_0|^2}{\rho c} \frac{1}{\pi kr}, \tag{9.60}$$

where Equations (9.52) and (9.53) have been used. The sound power radiated by a pulsating cylinder per unit length may easily be obtained by multiplying Equation (9.60) with $2\pi r$,

$$P_a' = \frac{2|A_0|^2}{k\rho c} = \frac{2|A_0|^2}{\rho \omega}. \tag{9.61}$$

For a cylinder at low frequencies, i.e., at $ka \ll 1$, the series expansions of the Hankel functions[3]

$$\lim_{x \to 0} H_0^{(2)}(x) = 1 - j\frac{2}{\pi}\ln(x), \tag{9.62}$$

and

$$\lim_{x \to 0} H_1^{(2)}(x) = j\frac{2}{\pi x}, \tag{9.63}$$

become useful (see Equations (C.6) and (C.7)). Equation (9.58) simplifies to

$$\lim_{ka \to 0} A_0 = \frac{1}{2} U_0 \rho c \pi ka. \tag{9.64}$$

In this limit Equations (9.55) and (9.57) reduce to

$$\lim_{ka \to 0} \hat{p}(r) = \frac{\pi}{2} U_0 \rho c k a H_0^{(2)}(kr) e^{j\omega t}, \tag{9.65}$$

and

$$\lim_{ka \to 0} \hat{u}_r(r) = -j\frac{\pi}{2} U_0 k a H_1^{(2)}(kr) e^{j\omega t}. \tag{9.66}$$

The near field of small cylindrical sources may be studied by taking Equations (9.65) and (9.66) to the limit $kr \to 0$. Equation (9.65) now becomes

$$\lim_{ka \to 0, r \to a} \hat{p}(r) = \frac{\pi}{2} U_0 \rho c k a \left(1 - j\frac{2}{\pi}\ln(kr)\right) e^{j\omega t}, \tag{9.67}$$

in which the imaginary term in the bracket evidently is much larger than the real part. The radial particle velocity becomes

$$\lim_{ka \to 0, r \to a} \hat{u}_r(r) = \frac{a}{r} U_0 e^{j\omega t}. \tag{9.68}$$

Hence, the sound pressure and the particle velocity are almost in quadrature close to a small pulsating cylinder.

The specific radiation impedance of a small cylinder at low frequencies is the ratio of Equation (9.67) to (9.68) when $r = a$,

$$\lim_{ka \to 0} Z_s = \rho c \left[\frac{\pi}{2}ka - jka\ln(ka)\right]. \tag{9.69}$$

Hence, the sound power radiated per unit length is

$$\lim_{ka \to 0} P_a' = \frac{1}{2}\text{Re}\{p(a)2\pi a U_0^*\} = \pi a |U_0|^2 \text{Re}\{Z_s\} = \frac{\pi^2}{2}\rho c k a^2 |U_0|^2, \tag{9.70}$$

---

[3] This result follows since $\frac{2}{\pi}\left(\ln\left(\frac{ka}{2}\right) + \gamma\right) = \frac{2}{\pi}(\ln(ka) - \ln(2) + \gamma) \simeq \frac{2}{\pi}\ln(ka)$, for $ka \ll 1$

which could also be found by inserting Equation (9.64) into Equation (9.61). Therefore, the sound power radiated by a small pulsating cylinder per unit length is proportional to the frequency at low frequencies.

A cylindrical source of first order is obtained by setting $m = \pm 1$ in Equation (9.54), which gives (with $k_r = k$),

$$\hat{p}(r, \varphi) = A_1^- H_1^{(2)}(kr) e^{j\varphi} e^{j\omega t} + A_1^+ H_1^{(2)}(kr) e^{-j\varphi} e^{j\omega t}, \tag{9.71}$$

where the two terms represent spinning modes in negative and positive φ directions respectively. However, spinning modes are often generated in pairs of equal magnitude, which allows the sound pressure to be written as,

$$\hat{p}(r, \varphi) = A_1 H_1^{(2)}(kr) \cos \varphi e^{j\omega t}. \tag{9.72}$$

Making use of Equation (9.56) gives expressions for the particle velocities in the radial and the circumferential directions,

$$\hat{u}_r(r, \varphi) = \frac{-jA_1}{\rho c} \frac{1}{kr} \left(H_1^{(2)}(kr) - kr H_0^{(2)}(kr) \right) \cos \varphi e^{j\omega t}, \tag{9.73}$$

and

$$\hat{u}_\varphi(r, \varphi) = \frac{-jA_1}{\rho c} \frac{1}{kr} H_1^{(2)}(kr) \sin \varphi e^{j\omega t}, \tag{9.74}$$

respectively.

For a source of radius $r = a$ the particle velocity in the radial direction has a $\cos \varphi$ dependence. Hence, this sound field can be produced by a cylinder that oscillates with the velocity U_1,

$$\hat{u}_r(a, \varphi) = U_1 \cos \varphi e^{j\omega t}, \tag{9.75}$$

which results in a relation between A_1 and U_1,

$$A_1 = \frac{j\rho c k a U_1}{H_1^{(2)}(ka) - ka H_0^{(2)}(ka)}, \tag{9.76}$$

which in the limit of low frequencies becomes,

$$\lim_{ka \to 0} A_1 = \frac{\pi}{2} \rho c (ka)^2 U_1. \tag{9.77}$$

Since the fluid is assumed to be inviscid no constraint on the tangential component of the particle velocity $\hat{u}_\varphi$ is imposed. It is also worth mentioning that the tangential particle velocity is in quadrature with the pressure at all distances to the source, which means that the tangential intensity is zero everywhere.

At low frequencies the expressions for the pressure and the particle velocity become,

$$\lim_{ka \to 0} \hat{p}(r, \varphi) = \frac{\pi}{2} \rho c (ka)^2 U_1 H_1^{(2)}(kr) \cos \varphi e^{j\omega t}, \tag{9.78}$$

$$\lim_{ka \to 0} \hat{u}_r(r, \varphi) = \frac{-j\pi}{2} U_1 \frac{(ka)^2}{kr} \left(H_1^{(2)}(kr) - kr H_0^{(2)}(kr) \right) \cos \varphi e^{j\omega t}, \tag{9.79}$$

and

$$\lim_{ka \to 0} \hat{u}_{\varphi}(r, \varphi) = \frac{-j\pi}{2} U_1 \frac{(ka)^2}{kr} H_1^{(2)}(kr) \sin \varphi e^{j\omega t}. \tag{9.80}$$

In the far field Equations (9.78) and (9.79) become,

$$\lim_{ka \to 0, r \to \infty} \hat{p}(r, \varphi) = j\frac{\pi}{2} \rho c (ka)^2 U_1 \sqrt{\frac{2}{\pi kr}} \cos \varphi e^{j(\omega t - kr + \pi/4)}, \tag{9.81}$$

and

$$\lim_{ka \to 0, r \to \infty} \hat{u}_r(r, \varphi) = \frac{j\pi}{2} U_1 (ka)^2 \sqrt{\frac{2}{\pi kr}} \cos \varphi e^{j(\omega t - kr + \pi/4)}, \tag{9.82}$$

which shows that the sound pressure is in phase with the particle velocity, and that the ratio of the pressure to the particle velocity is ρc. Hence, the sound power radiated per unit length from this source is

$$P'_a = \frac{1}{2}\rho c \int_0^{2\pi} |\hat{u}_r|^2 r d\varphi = \rho c \left(\frac{\pi}{2}\right)^2 k^3 a^4 |U_1|^2. \tag{9.83}$$

It can be seen that the oscillating cylinder is a weak source at low frequencies, since in this range the sound power radiated per unit length is proportional to the cube of the frequency.

Example 9.9 Consider a thin vibrating string of density ρ_s. The mechanical energy per unit length of this string when it vibrates with the amplitude U_1 is $E'_m = \frac{1}{2}\rho_s \pi a^2 |U_1|^2$. If the only losses are due to acoustic radiation, the vibration of the string will decay exponentially. At low frequencies the time constant becomes:

$$\tau = \frac{E'_m}{P'_a} = \frac{2}{\pi} \frac{\rho_s}{\rho c} \frac{1}{k^3 a^2}.$$

At low frequencies τ becomes very large, which indicates that for most musical instruments the losses due to direct radiation from the string are much smaller than other losses.

In the near field Equations (9.78), (9.79) and (9.80) become,

$$\lim_{ka \to 0, r \to a} \hat{p}(r, \varphi) = j\rho c \frac{ka^2}{r} U_1 \cos \varphi e^{j\omega t}, \tag{9.84}$$

$$\lim_{ka \to 0, r \to a} \hat{u}_r(r, \varphi) = U_1 \left(\frac{a}{r}\right)^2 \cos \varphi e^{j\omega t}, \tag{9.85}$$

and

$$\lim_{ka \to 0, r \to a} \hat{u}_{\varphi}(r, \varphi) = \left(\frac{a}{r}\right)^2 U_1 \sin \varphi e^{j\omega t}. \tag{9.86}$$

It is clear that the near field is very reactive with large amplitudes of the particle velocity, and that both the radial and the tangential particle velocity are in quadrature with the sound pressure.

For an arbitrary value of $m > 0$ Equation (9.54) gives

$$\hat{p}_m(r, \varphi) = A_m H_m^{(2)}(kr) e^{jm\varphi} e^{j\omega t}, \tag{9.87}$$

which may be interpreted as a wave spinning in the clockwise direction.

The particle velocities in the radial and circumferencial directions become

$$\hat{u}_{m,r}(r, \varphi) = \frac{jA_m}{2\rho c} \left(H_{m-1}^{(2)}(kr) - H_{m+1}^{(2)}(kr) \right) e^{jm\varphi} e^{j\omega t}, \tag{9.88}$$

and

$$\hat{u}_{m,\varphi}(r, \varphi) = \frac{-A_m}{\rho c} \frac{m}{kr} H_m^{(2)}(kr) e^{jm\varphi} e^{j\omega t}, \tag{9.89}$$

respectively.

In the far field Equation (9.53) leads to an expression for the specific impedance,

$$\lim_{r \to \infty} Z_s = \lim_{r \to \infty} \frac{\hat{p}_m(r, \varphi)}{\hat{u}_{m,r}(r, \varphi)} = -j\rho c \lim_{r \to \infty} \left(\frac{H_m^{(2)}(kr)}{\frac{1}{2} \left(H_{m-1}^{(2)}(kr) - H_{m+1}^{(2)}(kr) \right)} \right)$$

$$= -j\rho c \frac{e^{jm\pi/2}}{\frac{1}{2}(e^{j(m-1)\pi/2} - e^{j(m+1)\pi/2})} = \rho c, \tag{9.90}$$

which shows that in the far field, the sound field is locally a plane wave for any m.

This result may be used to obtain the radial intensity I_r in the far field

$$\lim_{r \to \infty} I_r = \frac{1}{2} \frac{1}{Z_s} |\hat{p}_m|^2 = \frac{1}{\rho c} \frac{|A_m|^2}{\pi kr}, \tag{9.91}$$

where Equation (9.53) has been used to obtain the last result.

If the surface of the cylinder $r = a$ moves with the radial velocity $\hat{u}_{m,r}(a, \varphi) = U_m e^{jm\varphi} e^{j\omega t}$ a relation between U_m and A_m may be found

$$\hat{u}_{m,r}(a, \varphi) = U_m e^{jm\varphi} e^{j\omega t}$$

$$= j \frac{A_m}{\rho c} \frac{dH_m^{(2)}(kr)}{d(kr)} \bigg|_{r=a} e^{jm\varphi} e^{j\omega t}. \tag{9.92}$$

In the limit of low frequencies, i.e., for $ka \ll 1$, Equation (C.15) leads to

$$U_m = \frac{A_m m!}{2\pi} \frac{1}{\rho c} \left(\frac{2}{ka} \right)^{m+1}, \tag{9.93}$$

or

$$A_m = \frac{2\pi \rho c}{m!} \left(\frac{ka}{2} \right)^{m+1} U_m. \tag{9.94}$$

Inserting into Equation (9.91) and reducing the expression results in

$$I_{m,r} = \frac{\pi \rho c}{(m!)^2} \left(\frac{ka}{2}\right)^{2m+1} \frac{2a}{r} |U_m|^2, \tag{9.95}$$

which is independent of φ.

The radiated sound power per unit length becomes

$$P'_{a,m} = 2\pi r I_r = \rho c \frac{4\pi^2 a}{(m!)^2} \left(\frac{ka}{2}\right)^{2m+1} |U_m|^2, \tag{9.96}$$

which shows that a high order source is a very poor radiator of sound at low frequencies.

Example 9.10 A standing wave of order m can be obtained by combining two spinning waves of orders $-m$ and m. Setting $U_{-m} = U_{m,s}/2$ and $U_m = U_{m,s}/2$ results in the radial velocity $\hat{u}_{m,r}(a, \varphi) = U_{m,s} \cos(m\varphi)e^{j\omega t}$ (index s refers to a standing wave). Due to the orthogonality of the circumferential terms the radiated powers may be summed. Making use of the fact that because of symmetry the power due to the term $-m$ equals the power due to the term m results in

$$P'_{a,m,s} = P'_{a,m} + P'_{a,-m}$$

$$= 2\rho c \frac{4\pi^2 a}{(m!)^2} \left(\frac{ka}{2}\right)^{2m+1} \left|\frac{U_{m,s}}{2}\right|^2$$

$$= \rho c \frac{8\pi^2 a}{(m!)^2} \left(\frac{ka}{2}\right)^{2m+1} \left|\frac{U_{m,s}}{2}\right|^2,$$

which reduces to Equation (9.83) for $m = 1$

Figure 9.9 shows the sound pressure levels generated by infinitely long cylindrical sources of different order with no variation in the longitudinal direction as functions of the distance. It is apparent that the first part of the decay becomes steeper the higher the order of the source; this region with a steep decay is the near field, and as can be seen the extent of the near field increases with the order of the source. Sufficiently far away the sound pressure decreases by the square root of the distance (3 dB when the distance is doubled or 10 dB per decade) for any source order – this defines the far field.

Figure 9.10 shows the normalised ratio of the particle velocity to the corresponding sound pressure as a function of the distance to cylindrical sources of different order. Near the source this ratio can be seen to depend on the order of the source; the higher the order the larger the ratio, indicating increasingly inefficient radiation; but sufficiently far away the ratio approaches unity (corresponding to the local wave impedance being identical with the characteristic impedance of air) irrespective of the order of the source. Since this was one of our boundary conditions it should not come as a surprise.

Finally Figure 9.11 shows the phase angle between the particle velocity and the sound pressure. These two quantities are in quadrature in a region near the source, and the size

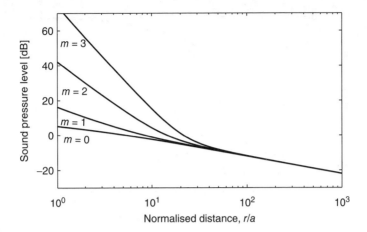

Figure 9.9 Amplitude of the sound pressure on the axis of symmetry of a cylindrical source ($\varphi = 0$) as a function of r from a to $1000a$ for $m = 0$, 1, 2 and 3 at $ka = 0.1$

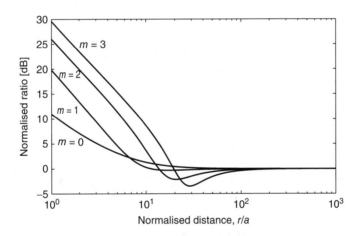

Figure 9.10 Ratio of the amplitude of the radial component of the particle velocity multiplied by ρc to the amplitude of the sound pressure on an axis of symmetry of a cylindrical source ($\varphi = 0$) as a function of r from a to $1000a$ for $m = 0$, 1, 2 and 3 at $ka = 0.1$

of the region increases with the order of the source. However, sufficiently far away the two quantities are in phase corresponding to a real-valued wave impedance, in agreement with the Sommerfeld radiation condition.

For a cylindrical source the general expression for the sound pressure is an infinite sum of terms (see Equation (9.54)),

$$\hat{p}(r, \varphi) = \sum_{m=-\infty}^{\infty} A_m H_m^{(2)}(kr) e^{jm\varphi} e^{j\omega t}. \tag{9.97}$$

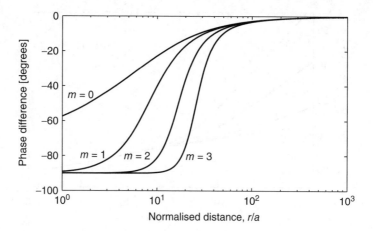

Figure 9.11 Phase angle between the radial component of the particle velocity and the sound pressure on the axis of symmetry of a cylindrical source ($\varphi = 0$) as a function of r from a to $1000a$ for $m = 0, 1, 2$ and 3 at $ka = 0.1$

The radial particle velocity is,

$$\hat{u}_r(r, \varphi) = \sum_{m=-\infty}^{\infty} \frac{jA_m}{\rho c} \frac{dH_m^{(2)}(kr)}{d(kr)} e^{jm\varphi} e^{j\omega t}. \tag{9.98}$$

On the surface of the cylinder this expression becomes,

$$\hat{u}_r(a, \varphi) = \sum_{m=-\infty}^{\infty} \frac{jA_m}{\rho c} \frac{dH_m^{(2)}(kr)}{d(kr)}\bigg|_{r=a} e^{jm\varphi} e^{j\omega t} = \sum_{m=-\infty}^{\infty} U_m e^{jm\varphi} e^{j\omega t}. \tag{9.99}$$

If the surface vibration of the source is $U(\varphi)e^{j\omega t}$ it is easy to show that the following Fourier pair establishes the relation between U_m and $U(\varphi)$,

$$U(\varphi) = \sum_{m=-\infty}^{\infty} U_m e^{jm\varphi}, \tag{9.100}$$

and

$$U_m = \frac{1}{2\pi} \int_0^{2\pi} U(\varphi)e^{-jm\varphi} d\varphi. \tag{9.101}$$

The sound pressure of a cylindrical source with a given surface velocity may be found by combining Equations (9.97), (9.99) and (9.101).

An infinite line-source placed on the surface of a rigid cylinder is the limiting case of a vibrating strip occupying an angle of Δ as Δ tends to zero. Hence, the surface velocity of the cylinder is:

$$U(\varphi) = \begin{cases} U & \text{if } -\Delta/2 < \varphi < \Delta/2 \\ 0 & \text{elsewhere.} \end{cases} \tag{9.102}$$

Inserting in Equation (9.101) yields:

$$U_m = \frac{1}{2\pi} \int_{-\Delta/2}^{\Delta/2} U e^{-jm\varphi} d\varphi, \tag{9.103}$$

and in the limit of $\Delta \to 0$, $U_m = \frac{U\Delta}{2\pi}$ is obtained for all m.

Using Equations (C.12) and (C.13) (in their Hankel function versions) along with Equation (9.99) results in

$$A_m = \begin{cases} j\rho c \frac{U\Delta}{2\pi} \frac{1}{H_1(ka)} & m = 0 \\ j\rho c \frac{U\Delta}{2\pi} \frac{2}{H_{m+1}(ka) - H_{m-1}(ka)} & m \neq 0. \end{cases} \tag{9.104}$$

Inserting Equation (9.104) into Equation (9.97) and making use of the fact that $A_{-m} = (-1)^m A_m$ along with the Hankel function version of Equation (C.10) results in

$$\hat{p}(r, \varphi) = j\rho c \frac{U\Delta}{2\pi} \left(\frac{H_0^{(2)}(kr)}{H_1^{(2)}(ka)} + \sum_{m=1}^{\infty} \frac{4 H_m^{(2)}(kr) \cos \varphi}{H_{m+1}^{(2)}(ka) - H_{m-1}^{(2)}(ka)} \right) e^{j\omega t}. \tag{9.105}$$

Normalising Equation (9.105) by the sound field produced by a small cylindrical source of zeroth order and with the velocity $U_0 = \frac{u\Delta}{2\pi}$ (see Equation (9.65)) gives

$$\frac{\hat{p}(r, \varphi)}{\hat{p}_0(r)} = \frac{2j}{\pi ka} \left(\frac{1}{H_1^{(2)}(ka)} + \frac{1}{H_0^{(2)}(kr)} \sum_{m=1}^{\infty} \frac{4 H_m^{(2)}(kr) \cos \varphi}{H_{m+1}^{(2)}(ka) - H_{m-1}^{(2)}(ka)} \right). \tag{9.106}$$

Figure 9.12 shows Equation (9.106) in the far field ($r \gg a$) for several values of ka. It is evident that at low frequencies, a line source on a cylinder is almost omni-directional – the radiated sound pressure of the source is approximately equal to the radiated sound pressure of a zeroth order source of the same strength. At higher frequencies the presence of the cylinder becomes important leading to an increased directivity of the response.

9.3.2 Scattering by Cylinders

It is possible to rewrite a scattering problem to an equivalent radiation problem. The approach makes use of linearity, which allows the total field outside a cylinder to be expressed as a sum of an incoming and a scattered field:

$$p_{\text{tot}} = p_{\text{inc}} + p_{\text{sc}}. \tag{9.107}$$

The incoming wave is the sound field in the absence of the cylinder – i.e., the undisturbed wave. It is worth noting that the incoming wave does not necessarily satisfy the Sommerfeld radiation condition, as is the case with a plane incoming wave. However, the scattered field, which represents the disturbance due to the cylinder, must satisfy the Sommerfeld radiation condition – the disturbance must 'wear off' at large distances. Therefore, the scattered field can be expressed as the sum given in Equation (9.97).

$$\hat{p}(r, \varphi) = \sum_{m=-\infty}^{\infty} A_m H_m^{(2)}(kr) e^{jm\varphi} e^{j\omega t}. \tag{9.108}$$

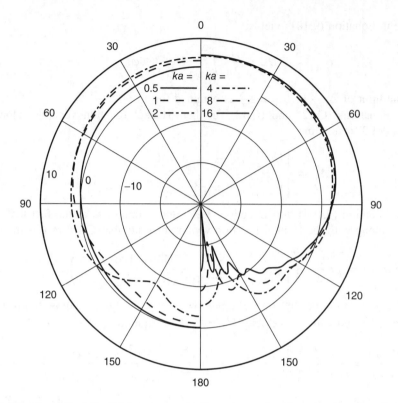

Figure 9.12 Normalised far-field pressure of a line source placed at $\varphi = 0$ on a cylinder at six frequencies

Hence, the scattered field corresponds to a radiation problem with a surface velocity in the radial direction given by Equation (9.98). If the scattered field can be constructed so that the total field satisfies the prescribed boundary condition on the surface of the scattering object, the total field must be the solution to the scattering problem due to the existence and uniqueness theorem [1].

In the following scattering of a normalised plane wave from a rigid infinite cylinder of radius a is considered. The boundary condition that must be fulfilled by the total field, is that the radial component of the velocity must vanish on the surface of the cylinder. Hence, the radial component of the scattered field must be equal in magnitude but opposite in phase to the incoming field,

$$\hat{u}_{sc,r}(a, \varphi) = -\hat{u}_{inc,r}(a, \varphi). \tag{9.109}$$

Without loss of generality the coordinate system can be aligned so that the plane wave travels along the negative x-axis in a Cartesian system. With this choice the problem is symmetric with respect to the x-axis, which has the consequence that the angular exponentials will appear in pairs of equal magnitude (see also Example 9.10) so that

Equations (9.97) and (9.98) specialise to

$$\hat{p}_{sc}(r, \varphi) = \sum_{m=0}^{\infty} A_m H_m^{(2)}(kr) \cos(m\varphi) e^{j\omega t}, \tag{9.110}$$

and

$$\hat{u}_{sc,r}(r, \varphi) = \sum_{m=0}^{\infty} \frac{jA_m}{\rho c} \frac{dH_m^{(2)}(kr)}{d(kr)} \cos(m\varphi) e^{j\omega t}. \tag{9.111}$$

In order to deal with the boundary condition on the surface of the cylinder, the incoming plane wave must be expressed in cylindrical coordinates. It is advantageous to express the incoming wave as a sum of Bessel functions (see Example 9.7), since the Neumann functions diverge for small arguments. The expansion must exist for all arguments, and it can immediately be concluded that only the expansion coefficients of the Bessel function are non-zero

$$e^{jkx} = e^{jkr \cos \varphi} = \sum_{m=0}^{\infty} B_m J_m(kr) \cos(m\varphi). \tag{9.112}$$

The expansion coefficients are found by multiplying Equation (9.112) with $\cos(n\varphi)$ and integrating over φ

$$\int_0^{2\pi} e^{jkr \cos \varphi} \cos(n\varphi) d\varphi = \int_0^{2\pi} \sum_{m=0}^{\infty} B_m J_m(kr) \cos(m\varphi) \cos(n\varphi) d\varphi. \tag{9.113}$$

The left-hand side is rewritten using Equation (C.16), and as a consequence of the orthogonality of the trigonometric functions only one term ($n = m$) of the sum on the right-hand side remains. Therefore,

$$B_m = \varepsilon_m j^m \tag{9.114}$$

is found, where $\varepsilon_m = 1$ for $m = 0$ and $\varepsilon_m = 2$ for $m > 0$.

The radial component of the particle velocity of the plane wave can now be expressed in terms of cylindrical components,

$$\hat{u}_{inc,r}(r, \varphi) = \sum_{m=0}^{\infty} \frac{jB_m}{\rho c} \frac{dJ_m(kr)}{d(kr)} \cos(m\varphi) e^{j\omega t} = \sum_{m=0}^{\infty} \frac{j^{m+1}\varepsilon_m}{\rho c} \frac{dJ_m(kr)}{d(kr)} \cos(m\varphi) e^{j\omega t}. \tag{9.115}$$

Due to orthogonality Equation (9.109) can be enforced term by term leading to an expression for A_m

$$A_m = - \varepsilon_m j^m \frac{dJ_m(kr)}{d(kr)}\bigg|_{r=a} \frac{dH_m^{(2)}(kr)}{d(kr)}\bigg|_{r=a}$$

$$= - \varepsilon_m j^m \frac{J_{m-1}(ka) - J_{m+1}(ka)}{H_{m-1}^{(2)}(ka) - H_{m+1}^{(2)}(ka)}, \tag{9.116}$$

where Equation (C.12) has been used in order to obtain the last result.

The total sound pressure becomes

$$\hat{p}(r, \varphi) = \sum_{m=0}^{\infty} (B_m J_m(kr) + A_m H_m^{(2)}(kr)) \cos(m\varphi) e^{j\omega t}, \tag{9.117}$$

and a non-normalised expression can be found by simply multiplying Equation (9.117) with the amplitude of the incoming wave.

Due to reciprocity Figure 9.12 can also be interpreted as the relative amplitude of the sound pressure on the surface of a cylinder when exposed to a plane wave coming from $0°$ degrees.

The sound field in the vicinity of the cylinder can be seen on Figure 9.13 for a few frequencies. The disturbance is modest at low frequencies, at high frequencies zones of shadow appear and a standing wave builds up in front of the cylinder.

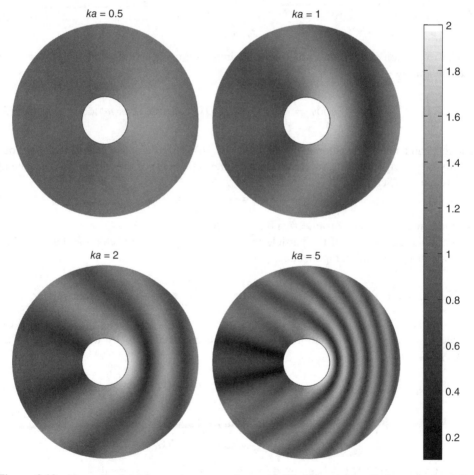

Figure 9.13 Normalised total sound pressure for scattering by a rigid cylinder, calculated for various values of ka. Sound incidence from the right

9.4 Spherical Waves

9.4.1 Radiation from Spherical Sources

One of the simplest examples of a source of finite extent to deal with analytically is that of a spherical source – and spherical sources are fairly good approximations to many real radiators of sound. We shall simplify the considerations by studying only spheres that vibrate symmetrically around the polar axis; see Figure 9.14. Under this condition, that is, with no variation with the azimuth angle φ, the Helmholtz equation in the spherical coordinate system becomes (cf. Equation (2.20c))

$$\frac{\partial^2 \hat{p}}{\partial r^2} + \frac{2}{r}\frac{\partial \hat{p}}{\partial r} + \frac{1}{r^2}\frac{\partial^2 \hat{p}}{\partial \theta^2} + \frac{1}{r^2 \tan\theta}\frac{\partial \hat{p}}{\partial \theta} + k^2 \hat{p} = 0. \tag{9.118}$$

To solve this equation it is now assumed that the sound pressure can be expressed as the product of functions that depend only on one coordinate:

$$\hat{p}(r, \theta, t) = p_r(r)p_\theta(\theta)e^{j\omega t}. \tag{9.119}$$

Insertion gives

$$\frac{r^2}{p_r(r)}\frac{d^2 p_r(r)}{dr^2} + \frac{2r}{p_r(r)}\frac{d\hat{p}_r(r)}{dr} + k^2 r^2 + \frac{1}{p_\theta(\theta)}\frac{d^2 p_\theta(\theta)}{d\theta^2} + \frac{1}{p_\theta(\theta)\tan\theta}\frac{dp_\theta(\theta)}{d\theta} = 0. \tag{9.120}$$

It is apparent that the first three terms on the left-hand side are functions only of r, and that the last two terms are functions only of θ. This leads to the conclusion that the sum of the first three terms must be a constant, C, whereas the sum of the last two terms must equal $-C$. Accordingly,

$$\frac{d^2 p_\theta(\theta)}{d\theta^2} + \frac{1}{\tan\theta}\frac{dp_\theta(\theta)}{d\theta} + Cp_\theta(\theta) = 0. \tag{9.121}$$

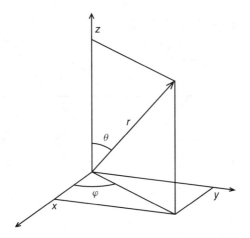

Figure 9.14   Spherical coordinate system

It can be shown that the solutions to this differential equation are finite only for certain discrete values of C,

$$C = m(m + 1), m = 0, 1, 2, 3, 4, \ldots \tag{9.122}$$

The corresponding solutions can be expressed in terms of a power series in $\cos\theta$, $P_m(\cos\theta)$. These functions are called Legendre functions of order m (see Appendix C.2). Figure 9.15 shows some examples of these functions. Note that the functions are normalised so that $P(\cos 0) = 1$.

The Legendre functions are orthogonal, that is,

$$\int_0^\pi P_m(\cos\theta)P_n(\cos\theta)\sin\theta \, \mathrm{d}\theta = \frac{2\delta_{mn}}{2m + 1}. \tag{9.123}$$

where δ_{mn} is the Kronecker symbol. As we shall see later, the orthogonality makes it possible to expand any function of θ into a set of such functions.

For a given value of m the function $p_r(r)$ must satisfy the equation

$$\frac{\mathrm{d}^2 p_r(r)}{\mathrm{d}r^2} + \frac{2}{r}\frac{\mathrm{d}p_r(r)}{\mathrm{d}r} + p_r(r)\left(k^2 - \frac{m(m + 1)}{r^2}\right) = 0. \tag{9.124}$$

The general solution to this differential equation is an arbitrary combination of a spherical Bessel function of order m and a spherical Neumann function of order m,

$$p_r(r) = A_m j_m(kr) + B_m n_m(kr). \tag{9.125}$$

These functions, which are closely related to the ordinary Bessel and Neumann function J_m and N_m, can be expressed in terms of the trigonometrical functions cos and sin; see Appendix C.3. Some spherical Bessel and Neumann functions are shown in Figure 9.16.

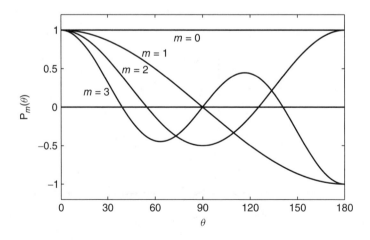

Figure 9.15 Legendre functions of order 0, 1, 2 and 3

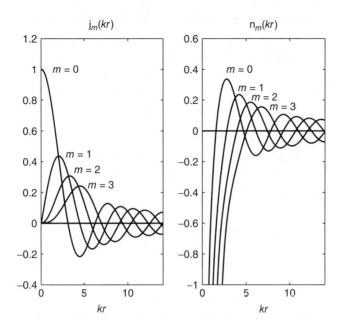

Figure 9.16 Spherical Bessel and Neumann functions for $m = 0, 1, 2$ and 3

However, $p_r(r)$ must also satisfy the boundary conditions. The boundary condition at infinity is known as the Sommerfeld radiation condition [1],

$$r\left(jk\hat{p}(r) + \frac{\partial\hat{p}(r)}{\partial r}\right) = jkr(\hat{p}(r) - \rho c\hat{u}_r(r)) \to 0 \quad \text{for} \quad kr \to \infty. \tag{9.126}$$

This condition expresses the fact that the sound field under free field conditions sufficiently far from a source of finite extent decays as $1/r$ and is locally plane, so that

$$\hat{u}_r(r) \simeq \frac{\hat{p}(r)}{\rho c}. \tag{9.127}$$

It can be shown that this is satisfied by Equation (9.119) if $B_m = -jA_m$. The combination of a spherical Bessel and Neumann function,

$$h_m^{(2)}(kr) = j_m(kr) - jn_m(kr). \tag{9.128}$$

is called a spherical Hankel function of the second kind.[4]

The resulting sound field can be written as a sum,

$$\hat{p}(r, \theta, t) = \sum_{m=0}^{\infty} A_m h_m^{(2)}(kr) P_m(\cos(\theta)) e^{j\omega t}. \tag{9.129}$$

The coefficients A_m depend on the vibrational pattern of the spherical source.

---

[4] With the alternative $e^{-i\omega t}$ sign convention the Sommerfeld radiation condition leads to the conclusion that $B_m = jA_m$. The corresponding combination of a spherical Bessel and Neumann function is a Hankel function of the *first* kind.

Example 9.11 Equations (9.126) and (9.127) ensure that the sound waves are travelling away from the source. The opposite case, that of a spherical converging sound field, is obtained by changing the spherical Hankel functions from the second kind to the first kind.[5] However, a converging spherical sound field is a rare phenomenon.

Example 9.12 All the Neumann functions diverge for $r \to 0$, from which it follows that $B_m = 0$ for interior problems. Therefore the sound field in a spherical cavity has the general form

$$\hat{p}(r, \theta, t) = \sum_{m=0}^{\infty} A_m j_m(kr) P_m(\cos \theta) e^{j\omega t}.$$

This may also be regarded as the interference between a diverging and a converging spherical sound field.

The lowest natural frequency in a spherical cavity with hard walls is associated with $m = 0$ and corresponds to the mode shape $\psi(r) = \sin(kr)/(kr)$.

We will now study the first few terms in the series separately. The first term is

$$\hat{p}(r, \theta) = A_0 h_0^{(2)}(kr) P_0(\cos \theta) e^{j\omega t}. \tag{9.130}$$

The spherical Hankel function of the second kind and order zero is fairly simple,

$$h_0^{(2)}(kr) = \frac{\sin kr}{kr} + j \frac{\cos kr}{kr} = \frac{j e^{-jkr}}{kr}. \tag{9.131}$$

The corresponding Legendre function is even simpler,

$$P_0(\cos \theta) = 1. \tag{9.132}$$

Inserting gives the sound pressure

$$\hat{p}(r) = j A_0 \frac{e^{j(\omega t - kr)}}{kr}, \tag{9.133}$$

from which we can compute the radial component of the particle velocity,

$$\hat{u}_r(r) = -\frac{1}{j\omega\rho} \frac{\partial \hat{p}(r)}{\partial r} = \frac{j A_0 e^{j(\omega t - kr)}}{\rho\omega r} \left(1 + \frac{1}{jkr}\right)$$

$$= \frac{\hat{p}(r)}{\rho c} \left(1 + \frac{1}{jkr}\right) \tag{9.134}$$

---

[5] With the alternative $e^{-j\omega t}$ sign convention a converging sound field is described by a sum of spherical Hankel functions of the *second* kind.

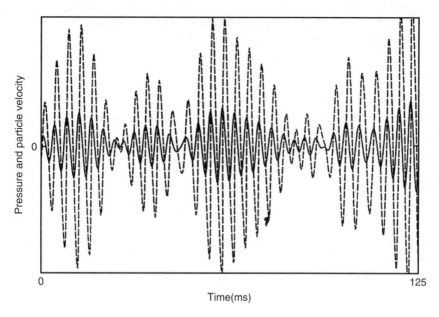

Figure 9.17   Sound pressure (–) and particle velocity multiplied by ρc (- -) close to a loudspeaker driven with one-third octave noise with a centre frequency of 250 Hz [10]

(cf. Equations (3.39) and (3.40)). Note the second term in the parenthesis, which has a phase shift of $-\pi/2$. This term becomes vanishingly small for $kr \to \infty$, in agreement with the boundary condition, Equation (9.126), but it is large at low frequencies, indicating the presence of a reactive near field. See Figure 9.17.

To generate this sound field the sphere must vibrate with a velocity that does not depend on θ, from which it follows that the boundary condition on the surface of the sphere is

$$\hat{u}_r(a) = U_0 e^{j\omega t}. \tag{9.135}$$

This is a *pulsating sphere*, that is, a sphere that expands and contracts harmonically. We can now express the constant A_0 in terms of the velocity of the sphere,

$$A_0 = -\frac{j\rho cka U_0 e^{jka}}{1 + \frac{1}{jka}}. \tag{9.136}$$

If we insert in Equation (9.133) we get the following expression for the sound pressure,

$$\hat{p}(r) = \frac{e^{jka}}{1 + jka} \frac{j\rho cka^2 U_0}{r} e^{j(\omega t - kr)}. \tag{9.137}$$

The sound intensity is

$$I_r(r) = \frac{1}{2} \text{Re}\{\hat{p}(r)\hat{u}_r^*(r)\} = \frac{\rho c}{2}\left(\frac{a}{r}\right)^2 |U_0|^2 \frac{(ka)^2}{1 + (ka)^2}. \tag{9.138}$$

It is worth noting that the sound intensity is related to the sound pressure in the same simple manner as in a plane propagating wave, that is,

$$I_r(r) = \frac{|\hat{p}(r)|^2}{2\rho c}. \tag{9.139}$$

This is a consequence of the fact that the near field term of the particle velocity is in quadrature with the sound pressure, which in this case has no near field term itself.[6]

The sound power radiated by the sphere can be obtained by integrating the normal component of the sound intensity over a surface that encloses the sphere, and since the sound intensity is constant on the spherical surface centred at the source we can simply multiply with the area,

$$P_a = I_r(r)4\pi r^2 = 2\pi a^2 \rho c |U_0|^2 \frac{(ka)^2}{1 + (ka)^2}. \tag{9.140}$$

Note that this is proportional to the square of the vibrational *acceleration* of the sphere at low frequencies.

The radiation impedance that loads the sphere is the ratio of the sound pressure to the volume velocity,

$$Z_{a,r} = \frac{\hat{p}(a)}{4\pi a^2 U_0 e^{j\omega t}} = \frac{jka}{1 + jka}\frac{\rho c}{4\pi a^2}. \tag{9.141}$$

If $ka \to 0$ the pulsating sphere becomes the point source we studied in Section 9.2, and Equations (9.137), (9.138) and (9.140) can be simplified as follows,

$$\hat{p}(r) = \frac{j\omega \rho Q e^{j(\omega t - kr)}}{4\pi r}, \tag{9.142}$$

$$I_r(r) = \frac{\rho c k^2 |Q|^2}{32\pi^2 r^2}, \tag{9.143}$$

$$P_a = \frac{\rho c k^2 |Q|^2}{8\pi}, \tag{9.144}$$

where

$$Q = 4\pi a^2 U_0 \tag{9.145}$$

is the volume velocity of the sphere. Under these conditions the imaginary part of the radiation impedance dominates completely,

$$Z_{a,r} \simeq j \text{Im}\{Z_{a,r}\} = \frac{jka\rho c}{4\pi a^2} = \frac{j\omega\rho}{4\pi a}, \tag{9.146}$$

and this is seen to be an acoustic mass,[7] indicating that a portion of the air around the sphere moves back and forth as if it were incompressible. The real part is

$$\text{Re}\{Z_{a,r}\} = \frac{\rho c k^2}{4\pi}. \tag{9.147}$$

---

[6] Terms that decay faster with the distance than $1/r$ are associated with the near field.

[7] Since $\text{Re}\{Z_{m,r}\} = (4\pi a^2)^2 \text{Re}\{Z_{a,r}\} = j\omega 4\pi a^3 \rho$, the imaginary part of the mechanical radiation impedance is seen to correspond to the mass of a sphere of air of radius $3^{1/3}a$.

Example 9.13 An alternative way of deriving Equation (9.144) makes use of the radiation impedance,

$$P_a = \frac{1}{2}\text{Re}\{\hat{p}Q^*e^{-j\omega t}\} = \frac{1}{2}|Q|^2\text{Re}\left\{\frac{\hat{p}}{Qe^{j\omega t}}\right\}$$

$$= \frac{1}{2}|Q|^2\text{Re}\{Z_{a,r}\} = \frac{\rho ck^2|Q|^2}{8\pi}.$$

This corresponds to integrating the normal component of the sound intensity over the surface of the sphere itself; cf. Equation (9.9).

Example 9.14 Since the sound power radiated by a small pulsating sphere is proportional to $(\omega Q)^2$ it follows that it should vibrate with a frequency independent volume *acceleration* in order to radiate the same sound power at all frequencies. Therefore, the vibrational displacement should be inversely proportional to the square of the frequency, indicating that very large displacements are required at low frequencies.

The second term in the general expression corresponds to a spherical source of first order,

$$\hat{p}(r,\theta) = A_1 h_1^{(2)}(kr)P_1(\cos\theta)e^{j\omega t}, \tag{9.148}$$

where the spherical Hankel function of the second kind and first order is

$$h_1^{(2)}(kr) = \frac{-e^{-jkr}}{kr}\left(1 + \frac{1}{jkr}\right), \tag{9.149}$$

and the Legendre function is

$$P_1(\cos\theta) = \cos\theta. \tag{9.150}$$

Inserting gives the sound pressure

$$\hat{p}(r,\theta) = -A_1\left(1 + \frac{1}{jkr}\right)\frac{e^{j(\omega t - kr)}}{kr}\cos\theta, \tag{9.151}$$

from which we can compute the radial component of the particle velocity,

$$\hat{u}_r(r,\theta) = -\frac{1}{j\omega\rho}\frac{\partial\hat{p}(r,\theta)}{\partial r} = -\frac{A_1}{\rho c}\left(1 - \frac{2}{(kr)^2} + \frac{2}{jkr}\right)\frac{e^{j(\omega t - kr)}}{kr}\cos\theta, \tag{9.152}$$

and the component in the θ-direction,

$$\hat{u}_\theta(r,\theta) = -\frac{1}{j\omega\rho}\frac{1}{r}\frac{\partial\hat{p}(r,\theta)}{\partial\theta} = \frac{jA_1}{\rho c}\left(1 + \frac{1}{jkr}\right)\frac{e^{j(\omega t - kr)}}{kr^2}\sin\theta. \tag{9.153}$$

Note that there are near field terms in the sound pressure and in both components of the particle velocity. The θ-component of the particle velocity is in fact exclusively associated with the near field and dies out faster with kr than the pressure or the radial component of the particle velocity.

The boundary condition on the surface of the sphere implies that the radial component of the particle velocity must agree with the normal component of the surface velocity of the sphere, and since it varies as $\cos\theta$ we can write

$$\hat{u}_r(a) = U_1 \cos\theta e^{j\omega t}, \tag{9.154}$$

from which it follows that

$$A_1 = -\frac{\rho c k a U_1 e^{jka}}{1 - \dfrac{2}{(ka)^2} + \dfrac{2}{jka}}, \tag{9.155}$$

and

$$\hat{p}(r,\theta) = \frac{\rho c k a U_1 e^{jka}}{1 - \dfrac{2}{(ka)^2} + \dfrac{2}{jka}} \left(1 + \frac{1}{jkr}\right) \frac{e^{j(\omega t - kr)}}{kr} \cos\theta. \tag{9.156}$$

A sphere that oscillates, that is, moves back and forth in the direction of the polar axis without changing its size or shape, will have a radial velocity given by Equation (9.154), since the velocity in the z-direction of an oscillating sphere is $U_1 e^{j\omega t}$ and since the vector product of the unit vector in the z-direction and the unit radial vector is $\cos\theta$.

The radial component of the sound intensity is

$$I_r(r,\theta) = \frac{1}{2}\text{Re}\{\hat{p}(r,\theta)\hat{u}_r^*(r,\theta)\} = \frac{|A_1|^2\cos^2\theta}{2\rho c(kr)^2}$$

$$= \frac{\rho c}{2}\left(\frac{a}{r}\right)^2 |U_1|^2 \cos^2\theta \frac{(ka)^4}{4 + (ka)^4}. \tag{9.157}$$

The sound pressure and the θ-component of the particle velocity are in quadrature, from which it follows that the θ-component of the intensity is identically zero,

$$I_\theta(r,\theta) = \frac{1}{2}\text{Re}\{\hat{p}(r,\theta)\hat{u}_\theta^*(r,\theta)\} = 0. \tag{9.158}$$

The sound power can be determined by integrating the radial component of the sound intensity over a spherical surface,

$$P_a = 2\pi \int_0^\pi I_r(r,\theta) r^2 \sin\theta d\theta = \pi a^2 \rho c |U_1|^2 \frac{(ka)^4}{4 + (ka)^4} \int_0^\pi \cos^2\theta \sin\theta d\theta$$

$$= \pi a^2 \rho c |U_1|^2 \frac{(ka)^4}{4 + (ka)^4} \int_{-1}^1 x^2 dx = \frac{\rho c}{3} 2\pi a^2 |U_1|^2 \frac{(ka)^4}{4 + (ka)^4}. \tag{9.159}$$

Note that the sound power of the vibrating sphere at low frequencies is proportional to the fourth power of the frequency, indicating very weak radiation.

If $ka \to 0$ Equations (9.156), (9.157) and (9.159) can be simplified as follows,

$$\hat{p}(r,\theta) = -\frac{\rho c (ka)^3 U_1 e^{j(\omega t - kr)}}{2kr}\left(1 + \frac{1}{jkr}\right)\cos\theta, \tag{9.160}$$

$$I_r(r,\theta) = \frac{\rho c (ka)^4}{8}|U_1|^2\left(\frac{a}{r}\right)^2\cos^2\theta, \tag{9.161}$$

$$P_a = \frac{\rho c (ka)^4}{6}\pi a^2 |U_1|^2. \tag{9.162}$$

There are obviously strong similarities between the sound field generated by a small vibrating sphere and the sound field produced by a point dipole, cf. Equations (9.12) and (9.13).

It takes a certain force to move the sphere back and forth. The ratio of this force to the resulting vibrational velocity is the mechanical radiation impedance of the sphere.[8] In the limit of $ka \to 0$ this ratio approaches the impedance of a mass,

$$\begin{aligned} Z_{m,r} &= \frac{\hat{F}}{U_1 e^{j\omega t}} = \frac{2\pi \int_0^\pi \hat{p}(a) a^2 \cos\theta \sin\theta \, d\theta}{U_1 e^{j\omega t}} \simeq j\rho c k \pi a^3 \int_{-1}^1 x^2 dx \\ &= j\omega\rho\frac{2\pi a^3}{3}, \end{aligned} \tag{9.163}$$

indicating that a portion of the air near the vibrating sphere is moving back and forth as if it were incompressible. There is a finite real part, of course, but it takes a better approximation to calculate the small radiation resistance,

$$\mathrm{Re}\{Z_{m,r}\} \simeq \rho c (ka)^4 \frac{\pi a^2}{3}. \tag{9.164}$$

Example 9.15 An alternative way of deriving Equation (9.162) makes use of the radiation impedance,

$$P_a = \frac{1}{2}\mathrm{Re}\{\hat{F}U_1^* e^{-j\omega t}\} = \frac{1}{2}|U_1|^2\mathrm{Re}\{Z_{m,r}\} \simeq \frac{\rho c (ka)^4 \pi a^2 |U_1|^2}{6};$$

cf. Example 9.13.

In the general case the sound pressure is a sum of many terms of the form

$$\hat{p}_m(r,\theta) = A_m h_m^{(2)}(kr) P_m(\cos\theta) e^{j\omega t}, \tag{9.165}$$

cf. Equation (9.129). The corresponding components of the particle velocity are

$$\hat{u}_{rm}(r,\theta) = -\frac{A_m}{j\omega\rho}\frac{dh_m^{(2)}(kr)}{dr}P_m(\cos\theta)e^{j\omega t} \tag{9.166}$$

---

[8] Since the volume velocity of the sphere is zero the *acoustic* radiation impedance of a spherical source of first order makes no sense.

and

$$\hat{u}_{\theta m}(r, \theta) = -\frac{A_m}{j\omega\rho} \frac{h_m^{(2)}(kr)}{r} \frac{dP_m(\cos\theta)}{d\theta} e^{j\omega t}.$$ (9.167)

Note that the θ-component of the particle velocity for a given value of m is in quadrature with the corresponding term of the expression for the sound pressure in the entire sound field.

From

$$\lim_{kr\to\infty} h_m^{(2)}(kr) = \frac{e^{-j(kr-\pi(m+1)/2)}}{kr}$$ (9.168)

(see Appendix C.3), we conclude that

$$\hat{p}_m(r, \theta) \to \frac{A_m}{kr} P_m(\cos\theta) e^{j(\omega t - kr + \pi(m+1)/2)} \quad \text{for} \quad kr \to \infty,$$ (9.169)

$$\hat{u}_{rm}(r, \theta) \to \frac{A_m}{\rho c k r} P_m(\cos\theta) e^{j(\omega t - kr + \pi(m+1)/2)} \quad \text{for} \quad kr \to \infty,$$ (9.170)

and

$$\hat{u}_{\theta m}(r, \theta) \to \frac{jA_m}{\rho c (kr)^2} \frac{dP_m(\cos\theta)}{d\theta} e^{j(\omega t - kr + \pi(m+1)/2)} \quad \text{for} \quad kr \to \infty.$$ (9.171)

It can be seen that the θ-component of the particle velocity is much smaller than the r-component in the far field, indicating that there is wave motion only in the r-direction. From Equations (9.169) and (9.170) we conclude that the r-component of the sound intensity is

$$I_{rm}(r) \to \frac{|A_m|^2}{2\rho c (kr)^2} P_m^2(\cos\theta) \quad \text{for} \quad kr \to \infty.$$ (9.172)

It is more difficult to show that the r-component of the sound intensity decays as $1/r^2$ also close to the sphere, but this is indeed the case, in agreement with the fact that the integral of this quantity over any spherical surface centred at the centre of the sphere is identical with the radiated sound power.

Figure 9.18 compares the sound pressure levels generated by axisymmetric spherical sources of different order on the axis of the sphere as functions of the distance. Note that the first part of the decay becomes steeper the higher the order of the source – the region with a steep decay defines the near field. However, sufficiently far away the sound pressure decreases by 6 dB when the distance is doubled for any source order (corresponding to 20 dB per decade); this defines the far field. The extent of the near field increases with the order of the source.

Figure 9.19 demonstrates the near and far field of spherical sources of different order in another way; the figure shows the normalised ratio of the particle velocity to the corresponding sound pressure. In the near field the ratio depends strongly on the order of the source; the higher the order the larger the ratio, indicating increasingly inefficient radiation for a given surface velocity; but sufficiently far away the ratio approaches unity (corresponding to the local wave impedance being identical with the characteristic impedance of air) irrespective of the order of the source, in agreement with the Sommerfeld radiation condition.

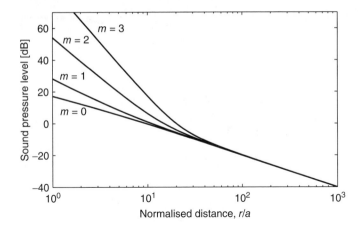

Figure 9.18 Amplitude of the sound pressure on the axis of symmetry of a spherical source ($\theta = 0$) as a function of r from a to $1000a$ for $m = 0$, 1, 2 and 3 at $ka = 0.1$

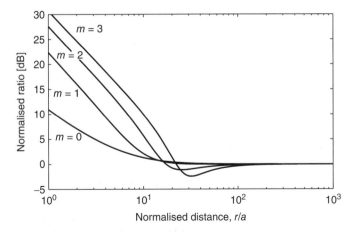

Figure 9.19 Ratio of the amplitude of the radial component of the particle velocity multiplied by ρc to the amplitude of the sound pressure on the axis of symmetry of a spherical source ($\theta = 0$) as a function of r from a to $1000a$ for $m = 0$, 1, 2 and 3 at $ka = 0.1$

Finally Figure 9.20 demonstrates the difference between the near field and the far field in yet another way. The figure shows the phase angle between the sound pressure and the particle velocity. If the source is small the two quantities are in quadrature in a region near the source (corresponding to a mass-like wave impedance); the size of this region increases with the order of the source. However, sufficiently far away the two quantities are in phase, again in agreement with the Sommerfeld radiation condition. If the source is large expressed in terms of the wavelength the wave impedance becomes predominantly real even close to the source.

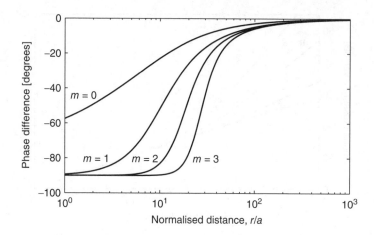

Figure 9.20 Phase angle between the radial component of the particle velocity and the sound pressure on the axis of symmetry of a spherical source ($\theta = 0$) as a function of r from a to $1000a$ for $m = 0, 1, 2$ and 3 at $ka = 0.1$

If the sphere is vibrating with a given, arbitrary axisymmetric velocity, $U(\theta)e^{j\omega t}$, the resulting sound field can be determined by expanding the velocity into *spherical harmonics*,

$$U(\theta) = \sum_{m=0}^{\infty} U_m P_m(\cos\theta). \tag{9.173}$$

The coefficients can be determined by multiplying both sides of this equation by $P_n(\cos\theta)$ and integrating over all angles,

$$\int_0^{\pi} U(\theta)P_n(\cos\theta)\sin\theta\,d\theta = \int_0^{\pi} \sum_{m=0}^{\infty} U_m P_m(\cos\theta)P_n(\cos\theta)\sin\theta\,d\theta$$
$$= \frac{1}{n+1/2}U_n, \tag{9.174}$$

where use has been made of the orthogonality of the Legendre functions (Equation (9.123)). It follows that

$$U_m = \left(m + \frac{1}{2}\right)\int_0^{\pi} U(\theta)P_m(\cos\theta)\sin\theta\,d\theta. \tag{9.175}$$

Each term in the expansion of $U(\theta)$ corresponds to a coefficient A_m in the general expression for the sound pressure, Equation (9.129). These coefficients are determined by equating with the r-component of the particle velocity at $r = a$,

$$A_m = -j\omega\rho U_m \left(\left.\frac{\mathrm{dh}_m^{(2)}(kr)}{dr}\right|_{r=a}\right)^{-1}. \tag{9.176}$$

It is instructive to study the case of a point source on a rigid sphere. This corresponds to the velocity

$$U(\theta) = \begin{cases} Q/(\pi(\Delta)^2) & \text{if } 0 < \theta < \Delta/a \\ 0 & \text{elsewhere,} \end{cases} \tag{9.177}$$

in the limit of $\Delta \to 0$. Inserting in Equation (9.175) gives

$$U_m = \lim_{\Delta \to 0} \left(m + \frac{1}{2}\right) \frac{Q}{\pi} \int_0^{\Delta/a} \frac{\sin\theta}{\Delta^2} P_m(\cos\theta) d\theta = (m + 1/2) \frac{Q}{2\pi a^2} \tag{9.178}$$

because $P_m(\cos 0) = 1$. It is apparent that it takes infinitely many spherical harmonics to describe a point source on a rigid sphere. Note also that

$$U_0 = \frac{Q}{4\pi a^2}, \tag{9.179}$$

in agreement with the fact that volume velocity of the zero order term of the expansion must equal the volume velocity of the point source. Combining Equations (9.129), (9.176) and (9.178) gives

$$\hat{p}(r, \theta) = -\frac{j\omega\rho Q e^{j\omega t}}{2\pi a^2} \sum_{m=0}^{\infty} (m + 1/2) h_m^{(2)}(kr) P_m(\cos(\theta)) \left(\frac{dh_m^{(2)}(kr)}{dr}\bigg|_{r=a}\right)^{-1}. \tag{9.180}$$

Figure 9.21 shows the sound pressure in the far field generated by a point source on a rigid sphere, normalised by the sound pressure generated by the point source in the absence of the sphere. As can be seen the sphere has very little influence at low frequencies. At higher frequencies the sphere gives rise to a certain directionality, and at high frequencies the sphere acts as an infinite, rigid wall, which doubles the sound pressure in the axial direction.

Because of the principle of reciprocity the curves shown in Figure 9.21 may also be interpreted as the sound pressure on the surface of a rigid sphere exposed to sound from a distant point source at (r, θ), normalised by the sound pressure at the origin in the absence of the sphere. At low frequencies the sphere has no influence on the sound pressure; at high frequencies the pressure on the sphere is doubled in the direction towards the source.

9.4.2 Scattering by Spheres

Scattering problems are often handled as equivalent radiation problems (cf. Section 9.3.2). The principle of superposition makes it possible to write the total sound pressure as the sum of the incident and scattered pressure,

$$p_{\text{tot}} = p_{\text{inc}} + p_{\text{sc}}. \tag{9.181}$$

The incident pressure is the sound pressure in absence of the object, whereas the scattered pressure represents the disturbance due to the object. At low frequencies the object does not disturb the sound field significantly. Hence, it is expected that the scattered sound pressure level is low compared with the incident field, so that the total pressure at low frequencies equals the incident pressure. (This provides a check of the equations derived in what follows.)

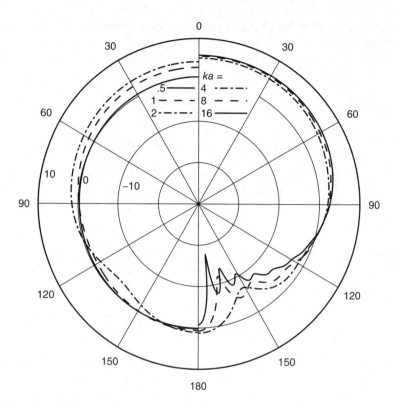

Figure 9.21 Normalised sound pressure in the far field generated by a point source on a rigid sphere, calculated for various values of ka

The scattered pressure must satisfy the Sommerfeld radiation condition (see Equation (9.126)) whereas this is not necessary for the incident pressure, which can be a plane wave for instance. The procedure makes use of the theorem of uniqueness [1]: If it is possible to construct the scattered pressure, so that the boundary condition at the surface of the scattering object is fulfilled for the total pressure or particle velocity, as well as the Sommerfeld radiation condition for the scattered pressure, then the uniqueness theorem ensures that the solution found is the correct and only solution.

Consider a plane wave travelling along the z-axis in its positive direction. The resulting sound field will then be axisymmetric so Equation (9.118) will describe the sound field. Hence, the scattered sound pressure can be written as a sum of outgoing waves (see Equation (9.129)),

$$\hat{p}_{sc}(r, \theta, t) = \sum_{m=0}^{\infty} A_m h_m^{(2)}(kr) P_m(\cos \theta) e^{j\omega t}, \tag{9.182}$$

and the particle velocity correspondingly,

$$\hat{u}_{sc,r}(r, \theta, t) = \frac{-1}{j\omega\rho} \sum_{m=0}^{\infty} A_m \frac{dh_m^{(2)}(kr)}{dr} P_m(\cos \theta) e^{j\omega t}. \tag{9.183}$$

For a rigid sphere the total radial particle velocity at the surface of the sphere $r = a$ must be zero. In order to establish the equations corresponding to this condition, the plane wave must be expressed in spherical coordinates,

$$
\begin{aligned}
\hat{p}_{\text{inc}} &= e^{-jkr\cos\theta} e^{j\omega t} \\
&= \sum_{m=0}^{\infty} B_m j_m(kr) P_m(\cos\theta) e^{j\omega t},
\end{aligned}
\tag{9.184}
$$

where use has been made of the fact that the coefficients of the Neumann functions must all be zero, since the plane wave is of finite amplitude everywhere, including the origin of the coordinate system. The expansion coefficients B_m are found by multiplying with P_n, integrating over the surface of the sphere and making use of the fact that Legendre functions are orthogonal. The calculations are not entirely trivial – see Appendix C.4. Thereby

$$
B_m = (2m + 1)(-j)^m
\tag{9.185}
$$

is found. The particle velocity in the radial direction due to the incoming wave is then

$$
\hat{u}_{\text{inc},r}(r, \theta, t) = \frac{-1}{j\omega\rho} \sum_{m=0}^{\infty} (2m + 1)(-j)^m P_m(\cos\theta) \frac{d j_m(kr)}{dr} e^{j\omega t}.
\tag{9.186}
$$

The boundary condition for the total particle velocity at $r = a$ is $\hat{u}_{\text{tot},r} = 0$, which implies that $\hat{u}_{\text{sc},r} = -\hat{u}_{\text{inc},r}$. By making use of Equations (9.183) and (9.186) term by term an expression for A_m is found,

$$
\begin{aligned}
A_m &= -(2m + 1)(-j)^m \left. \frac{d j_m(kr)}{d(kr)} \right|_{r=a} \Bigg/ \left. \frac{d h_m^{(2)}(kr)}{d(kr)} \right|_{r=a} \\
&= -(2m + 1)(-j)^m \frac{m j_{m-1}(ka) - (m + 1) j_{m+1}(ka)}{m h_{m-1}^{(2)}(ka) - (m + 1) h_{m+1}^{(2)}(ka)}
\end{aligned}
\tag{9.187}
$$

(see Appendix C.3). The total sound pressure becomes

$$
\hat{p}_{\text{tot}}(r, \theta, t) = \sum_{m=0}^{\infty} (B_m j_m(kr) + A_m h_m^{(2)}(kr)) P_m(\cos\theta) e^{j\omega t}.
\tag{9.188}
$$

Figure 9.22 shows the relative magnitude of the total sound pressure for a number of frequencies. Note that the sound field is practically undisturbed at low frequencies and that a standing wave builds up in front of the sphere at higher frequencies.

9.4.3 Ambisonics

Ambisonics is a reproduction technique based on spherical harmonics – but in principle limited to recreating the desired (recorded) sound field at the centre of a spherical loudspeaker array under free field conditions [11]. The first step of the ambisonics procedure involves recording (or encoding) the desired sound field with a special device. For example, a so-called sound field microphone combines one pressure microphone and

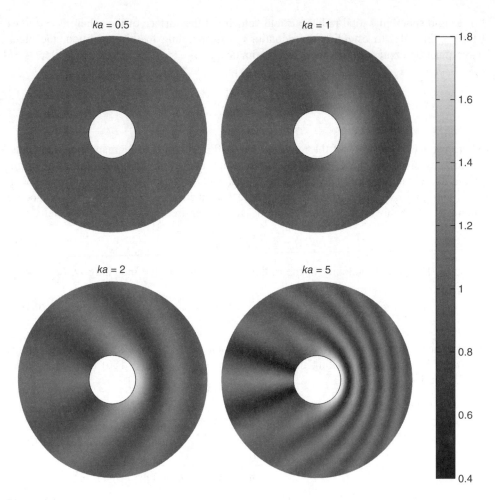

Figure 9.22   Normalised total sound pressure for scattering by a rigid sphere, calculated for various values of ka. Sound incidence from the right

three figure-of-eight microphones oriented perpendicularly; this provides the first four components of the expansion. A more general expansion can be provided by a spherical microphone array [12].

The expansion of the desired sound field into a (truncated) sum of three-dimensional orthogonal spherical functions (which is analogous to the expansion of a plane wave given by Equation (9.184)) makes it possible to establish a system of linear equations that can be solved for the loudspeaker signals [13].

9.5   Plane Sources

Many sources of sound are plane, as for instance vibrating panels. The mathematical treatment of a plane source is simple only if it is baffled, that is, if it is a part of an infinite, rigid plane surface. This is often a good approximation in a substantial part of the frequency range.

9.5.1 The Rayleigh Integral

An infinite plane surface at $z = 0$ is vibrating with the velocity $U(x, y)$. We divide it into small rectangles (dx', dy') and regard each rectangle as a point source on a rigid baffle, which generates the sound pressure

$$\hat{p}(x, y, z) = \frac{j\omega\rho \, U(x', y') \, e^{j(\omega t - kr)}}{2\pi \, r} \, dx'dy', \tag{9.189}$$

where

$$r = \sqrt{(x - x')^2 + (y - y')^2 + z^2} \tag{9.190}$$

is the distance between the observation point in front of the baffle and the point source. Note that the sound pressure has been doubled because of the image source due to the plane (cf. Example 9.3). The complete sound field is found by superposition, which leads to *Rayleigh's integral* [1] (or strictly, Rayleigh's second integral):

$$\hat{p}(x, y, z) = \frac{j\omega\rho}{2\pi} \iint_{-\infty}^{\infty} U(x', y') \frac{\exp\left(j(\omega t - k\sqrt{(x - x')^2 + (y - y')^2 + z^2})\right)}{\sqrt{(x - x')^2 + (y - y')^2 + z^2}} dx'dy'. \tag{9.191}$$

This can be expressed in more condensed form in terms of the free field Green's function from Example 9.1,

$$\hat{p}(\mathbf{r}) = 2 \, j\omega\rho \, e^{j\omega t} \iint_{-\infty}^{\infty} U(\mathbf{r}')G(\mathbf{r}, \mathbf{r}') \, dS'. \tag{9.192}$$

Note that Rayleigh's integral is a two-dimensional convolution of the vibrational velocity and the Green's function. The implications of this observation will become clear in Section 9.5.2.

The sound field generated by a *finite* vibrating panel mounted in an infinite, rigid baffle is obtained by integrating over the panel. In the far field we can ignore the weak dependence of the amplitude of the integrand on the position but not the dependence of the phase, and thus

$$\hat{p}(r, \theta, \varphi) = \frac{j\omega\rho \, e^{j(\omega t - kr)}}{2\pi r} \iint_{S} U(x', y') \, e^{j(kx' \sin\theta \cos\varphi + ky' \sin\theta \sin\varphi)} dx'dy', \tag{9.193}$$

where r, θ and φ are the coordinates of the receiver point expressed in a spherical coordinate system centred in the middle of the panel, and S is the area of the panel. Note that it has also been assumed that the observation point is so far away that the line between this point and the running integration point is at the same angle, θ, at all points on the panel. This expression can be written as the product of the time factor $e^{j\omega t}$, the radial factor e^{-jkr}/r, and the *directivity function* of the source $D(\theta, \varphi)$,

$$\hat{p}(r, \theta, \varphi) = \frac{e^{j(\omega t - kr)}}{r} D(\theta, \varphi), \tag{9.194}$$

where

$$D(\theta, \varphi) = \frac{j\omega\rho}{2\pi} \iint_S U(x', y') \, e^{j(kx' \sin\theta \cos\varphi + ky' \sin\theta \sin\varphi)} \, dx' dy'. \qquad (9.195)$$

Example 9.16 The sound pressure in the far field produced by a circular piston in an infinite baffle depends only on r and θ,

$$\hat{p}(r, \theta) = \frac{j\omega\rho U}{2\pi r} e^{j(\omega t - kr)} \int_0^{2\pi} \int_0^a e^{jkw \sin\theta \cos\kappa} w \, dw \, d\kappa,$$

where U is the velocity of the piston, and w and κ are the coordinates of the running point on the piston in a polar coordinate system. Since the (cylindrical) Bessel function of order zero, $J_0(z)$, can be written as

$$J_0(z) = \frac{1}{2\pi} \int_0^{2\pi} e^{jz \cos\beta} d\beta,$$

and the Bessel functions of order zero and one satisfy the following relation (see Equation (C.14)),

$$J_1(z) = \frac{1}{z} \int z J_0(z) dz,$$

the expression for the far field sound pressure becomes

$$\hat{p}(r, \theta) = \frac{j\omega\rho U}{r} \frac{a J_1(ka \sin\theta)}{k \sin\theta} e^{j(\omega t - kr)} = \frac{j\omega\rho Q e^{j(\omega t - kr)}}{2\pi r} \left[\frac{2 J_1(ka \sin\theta)}{ka \sin\theta} \right],$$

where the volume velocity of the piston, $Q = \pi a^2 U$, has been introduced. The function $2J_1(z)/z$ is plotted in Figure 9.23, and Figure 9.24 shows the directivity function of the circular piston as a function of the frequency, normalised to unity in the axial direction. Note that the far field sound pressure in the axial direction is identical with the sound pressure generated by a monopole on a rigid baffle with the same volume velocity as the piston at all frequencies. At low frequencies the piston radiates as a monopole on a baffle in all directions.

Example 9.17 The sound pressure anywhere on the axis of symmetry in front of a circular piston in an infinite baffle can be calculated from Equation (9.191) without any approximations. Expressed in cylindrical coordinates Rayleigh's integral becomes

$$\hat{p}(0, 0, z) = \frac{j\omega\rho U e^{j\omega t}}{2\pi} \int_0^{2\pi} \int_0^a \frac{w e^{-jk\sqrt{z^2 + w^2}}}{\sqrt{z^2 + w^2}} dw d\phi$$

$$= j\omega\rho U e^{j\omega t} \int_o^a \frac{w e^{-jk\sqrt{z^2 + w^2}}}{\sqrt{z^2 + w^2}} dw,$$

where (w, ϕ) is the position of the running integration point on the piston. This integral can be solved if we introduce the distance between the observation point on the axis and the running point,

$$h = \sqrt{z^2 + w^2},$$

and make use of the fact that

$$\frac{dh}{dw} = \frac{w}{\sqrt{z^2 + w^2}}.$$

The result is

$$\hat{p}(0, 0, z) = \mathrm{j}\omega\rho U \mathrm{e}^{\mathrm{j}\omega t} \int_z^{\sqrt{z^2+a^2}} \mathrm{e}^{-\mathrm{j}kh}\,dh = \rho c U\,(\mathrm{e}^{-\mathrm{j}kz} - \mathrm{e}^{-\mathrm{j}k\sqrt{z^2+a^2}})\mathrm{e}^{\mathrm{j}\omega t}$$

$$= 2\mathrm{j}\rho c U\,\mathrm{e}^{\mathrm{j}(\omega t - k(z+\Delta))}\,\sin(k\Delta),$$

where, in addition, the quantity

$$\Delta = \frac{\sqrt{z^2 + a^2} - z}{2}$$

has been introduced. The axial sound pressure is plotted in Figure 9.25. At large distances it approaches the far field expression derived in Example 9.1, as one would expect. Note the surprising fact that the pressure vanishes at some positions. This is due to destructive interference caused by the varying distance between the observation point and points on the piston. It is also possible to interpret the two complex exponentials that compose the sine in the resulting expression as the direct wave and a wave reflected from the edge of the piston; thus the fluctuations may be regarded as interference between these two waves.

Note that Equation (9.193) can be written as

$$\hat{p}(r, \theta, \varphi) = \frac{\mathrm{j}\omega\rho\mathrm{e}^{\mathrm{j}(\omega t - kr)}}{2\pi r} \iint_{-\infty}^{\infty} U(x', y')\mathrm{e}^{\mathrm{j}(k_x x' + k_y y')}\,dx'\,dy', \qquad (9.196)$$

where

$$(k_x, k_y) = (k\sin\theta\cos\varphi,\ k\sin\theta\sin\varphi), \qquad (9.197)$$

and U is zero outside of the panel. Comparing with familiar expression for a Fourier transform of the time function $x(t)$,

$$X(\omega) = \int_{-\infty}^{\infty} x(t)\mathrm{e}^{-\mathrm{j}\omega t}\,dt \qquad (9.198a)$$

and its inverse

$$x(t) = \frac{1}{2\pi} \int_{-\infty}^{\infty} X(\omega)\mathrm{e}^{\mathrm{j}\omega t}\,d\omega \qquad (9.198b)$$

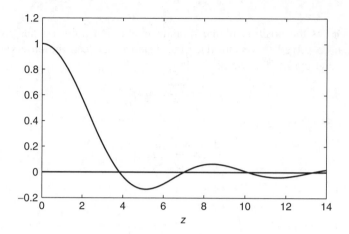

Figure 9.23 The function $2J_1(z)/z$

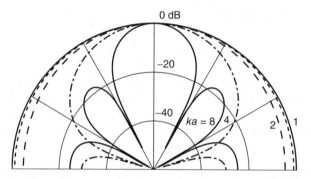

Figure 9.24 Directional pattern of a baffled circular piston for different values of the normalised frequency ka

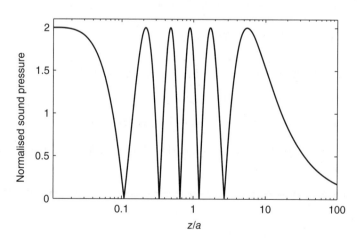

Figure 9.25 Sound pressure on the axis normalised by $\rho c U$ as a function of z/a for $ka = 35$

(cf. Equations (B2) and (B3)) leads to the conclusion that the expression for the sound pressure in the far field is proportional to the two-dimensional spatial Fourier transform of the velocity $U(x, y)$, that is,

$$\hat{p}(r, \theta, \varphi) = \frac{j\omega\rho e^{j(\omega t - kr)}}{2\pi r} W(k_x, k_y),\tag{9.199}$$

where

$$W(k_x, k_y) = \int_{-\infty}^{\infty} \int_{-\infty}^{\infty} U(x, y)e^{j(k_x x + k_y y)}dxdy,\tag{9.200a}$$

is the *wavenumber transform* of $U(x, y)$ and

$$U(x, y) = \frac{1}{4\pi^2} \int_{-\infty}^{\infty} \int_{-\infty}^{\infty} W(k_x, k_y)e^{-j(k_x x + k_y y)}dk_x dk_y,\tag{9.200b}$$

is the inverse transform.[9] The space-wavenumber pair, Equation (9.200), is analogous to the more familiar time-frequency pair, Equation (9.198).

Example 9.18 A baffled rectangular piston in the plane $z = 0$ is vibrating with the velocity

$$U(x, y)e^{j\omega t} = \begin{cases} Ue^{j\omega t} & \text{if } \frac{-L_x}{2} < x < \frac{L_x}{2} \text{ and } \frac{-L_y}{2} < y < \frac{L_y}{2} \\ 0 & \text{elsewhere.} \end{cases}$$

From Equation (9.196) we conclude that

$$\hat{p}(r, \theta, \varphi) = \frac{j\omega\rho U e^{j(\omega t - kr)}}{2\pi r} \int_{-\frac{L_x}{2}}^{\frac{L_x}{2}} \int_{-\frac{L_y}{2}}^{\frac{L_y}{2}} \exp(j(kx \sin\theta \cos\varphi + ky \sin\theta \sin\varphi))dxdy$$

$$= \frac{j\omega\rho L_x L_y U e^{j(\omega t - kr)}}{2\pi r} \frac{\sin\left(\frac{kL_x}{2}\sin\theta \cos\varphi\right)}{\frac{kL_x}{2}\sin\theta \cos\varphi} \frac{\sin\left(\frac{kL_y}{2}\sin\theta \sin\varphi\right)}{\frac{kL_y}{2}\sin\theta \sin\varphi}.$$

Note that the piston simply radiates as a baffled point source with the volume velocity $L_x L_y U$ at low frequencies.

9.5.2 The Wavenumber Approach

Let us look at the case where an infinite plate at $z = 0$ is vibrating with the velocity

$$U(x, y) e^{j\omega t} = Ae^{j(\omega t - k_x x)},\tag{9.201}$$

corresponding to a bending wave propagating in the x-direction with the speed

$$c_p = \frac{\omega}{k_x}.\tag{9.202}$$

---

[9] Note that the sign of the argument of the exponential in the Fourier transform has been reversed.

The wavelength of this wave is

$$\lambda_p = \frac{2\pi}{k_x}. \tag{9.203}$$

The resulting sound field must be independent of y and must depend on x in the same way as the wave on the plate. A possible expression for the sound pressure generated by the vibrations of the plate in the half space above it could be a plane wave of the form

$$\hat{p}(x, y, z) = p_0 e^{j(\omega t - k_x x - k_z z)}, \tag{9.204}$$

where k_x and k_z must satisfy the equation

$$k_z = \sqrt{k^2 - k_x^2}, \tag{9.205}$$

since, from the Helmholtz equation,

$$k_x^2 + k_y^2 + k_z^2 = k^2, \tag{9.206}$$

and since $k_y = 0$. The corresponding particle velocity in the direction normal to the plate is

$$\hat{u}_z(x, y, z) = \frac{-1}{j\omega\rho}\frac{\partial p}{\partial z} = \frac{p_0}{\rho c}\frac{k_z}{k}e^{j(\omega t - k_x x - k_z z)}, \tag{9.207}$$

and this must agree with the plate velocity at $z = 0$. This gives an expression for p_0, and we finally get

$$\hat{p}(x, y, z) = \frac{A\rho c k}{k_z}e^{j(\omega t - k_x x - k_z z)}. \tag{9.208}$$

The sound field generated by the bending wave on the plate depends significantly on whether the argument of the square root of Equation (9.205) is positive or negative. When $k > k_x$ the argument is positive and k_z is a real quantity. In this case Equation (9.208) describes a plane wave propagating in the direction indicated by the vector (k_x, k_z). The bending wave is said to be *supersonic*, because it propagates with a speed that exceeds the speed of sound in air. The trace of the plane wave on the plate propagates with the same speed,

$$c_p = \frac{c}{\sin\theta}, \tag{9.209}$$

where

$$\theta = \arctan(k_x/k_z), \tag{9.210}$$

the angle measured from the normal direction, indicates the direction of the plane wave; see Figure 9.26.

When $k < k_x$ the argument of the square root is negative, from which it follows that k_z is purely imaginary. An imaginary value of k_z corresponds to a wave that decays or grows exponentially with z. The latter is physically unreasonable (and disagrees with the Sommerfeld radiation condition), which leads to the conclusion that the sign of the square root must be chosen as follows,

$$k_z = -j\sqrt{k_x^2 - k^2}, \tag{9.211}$$

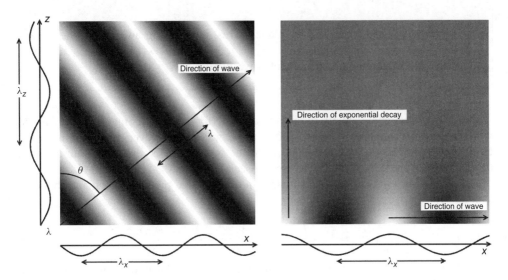

Figure 9.26 Sound field above a plate carrying a plane propagating flexural wave. In the left-hand figure the flexural wave is supersonic, and in the right-hand figure it is subsonic

when $k < k_x$. The decaying plane wave is an *evanescent wave*. It is easy to see from Equation (9.208) that the normal component of the particle velocity is in quadrature with the sound pressure in the evanescent wave, from which it follows that the sound intensity is zero. In other words, there is no radiation to the far field; only local wave motion close to the plate. This phenomenon, which is sometimes referred to as the hydrodynamic short circuit, occurs when the bending wave is *subsonic* (ie, travels slower than the speed of sound in air) and the wavelength in the plate is shorter than the wavelength in air; see Figure 9.26.

Example 9.19 Let an infinite plate at $z = 0$ vibrate as

$$U(x, y)e^{j\omega t} = A\cos(k_x x)\cos(k_y y)e^{j\omega t},$$

corresponding to a standing wave pattern in the plate (i.e., bending modes in an infinite plate). By writing the cosines as sums of exponentials the velocity can be decomposed into four propagating plane bending waves, which are dealt with as above. The resulting sound pressure is

$$\hat{p}(x, y, z) = \frac{A\rho c k}{k_z}\cos(k_x x)\cos(k_y y)e^{j(\omega t - k_z z)},$$

where

$$k_z = \sqrt{k^2 - k_x^2 - k_y^2}.$$

When

$$k^2 > k_x^2 + k_y^2$$

the plate modes gives rise to four plane waves, propagating away from the plate in the directions (k_x, k_y, k_z), $(k_x, -k_y, k_z)$, $(-k_x, k_y, k_z)$ and $(-k_x, -k_y, k_z)$; otherwise they generate an evanescent wave field.

A general vibrational velocity of an infinite plate (for example a finite panel in a baffle) will not generate a plane wave (or a few plane waves). However, the resulting sound field can nevertheless be decomposed into plane waves, that is, expanded into a system of plane and evanescent waves. This can be seen as follows.

Let us introduce the wavenumber transform of the sound pressure in a plane above the source, say $z = z_m$,

$$P(k_x, k_y, z_m) = \int_{-\infty}^{\infty} \int_{-\infty}^{\infty} \hat{p}(x, y, z_m) e^{j(k_x x + k_y y)} dx dy. \tag{9.212}$$

An inverse two-dimensional Fourier transform brings us back from the wavenumber domain to the spatial domain,

$$\hat{p}(x, y, z_m) = \frac{1}{4\pi^2} \int_{-\infty}^{\infty} \int_{-\infty}^{\infty} P(k_x, k_y, z_m) e^{-j(k_x x + k_y y)} dk_x dk_y. \tag{9.213}$$

Inspection of Equation (9.213) reveals that the wavenumber spectrum $P(k_x, k_y, z_m)$ represents a decomposition of the sound field into a distribution of plane and evanescent waves in the same way as $X(\omega)$ in Equation (9.198a) represents a decomposition of $x(t)$ into complex exponentials. This decomposition into plane/evanescent waves makes it possible to extrapolate the field: one simply follows the plane waves making use of the fact that

$$k_z = \sqrt{k^2 - k_x^2 - k_y^2}, \tag{9.214}$$

where the sign of the square root when its argument is negative must be chosen so as to give a sound field that decays exponentially with the distance to the source plane (cf. Equation (9.211)). Thus, if we have determined the wavenumber spectrum of the sound pressure in a given plane, it is a simple matter to extrapolate the sound field to another plane, say $z = z_p$; one simply multiplies with the 'propagator' $e^{-jk_z(z_p - z_m)}$ (which depends on k_x and k_y) and transforms the resulting wavenumber spectrum back to the spatial domain using Equation (9.213),

$$\hat{p}(x, y, z_p) = \frac{1}{4\pi^2} \int_{-\infty}^{\infty} \int_{-\infty}^{\infty} P(k_x, k_y, z_m) e^{-j(k_x x + k_y y + k_z(z_p - z_m))} dk_x dk_y. \tag{9.215}$$

One can also calculate the particle velocity. For example, the particle velocity component in the z-direction is

$$\hat{u}_z(x, y, z_p) = \frac{-1}{j\omega\rho} \frac{\partial \hat{p}(x, y, z)}{\partial z} \bigg|_{z=z_p}$$

$$= \frac{1}{4\pi^2 \rho c} \int\!\!\int_{-\infty}^{\infty} \frac{k_z}{k} P(k_x, k_y, z_m) e^{-j(k_x x + k_y y + k_z(z_p - z_m))} dk_x dk_y. \tag{9.216}$$

The normal component of the particle velocity must agree with the vibrational velocity of the source in the plane $z = 0$, that is,

$$U(x, y)e^{j\omega t} = \frac{1}{4\pi^2 \rho c} \int_{-\infty}^{\infty} \int_{-\infty}^{\infty} \frac{k_z}{k} P(k_x, k_y, 0)e^{-j(k_x x + k_y y)} dk_x dk_y. \tag{9.217}$$

Comparing with Equation (9.200) shows that

$$P(k_x, k_y, 0) = \frac{k\rho c}{k_z} W(k_x, k_y), \tag{9.218}$$

which, with Equation (9.215) leads to the conclusion that the sound pressure at an arbitrary position in front of the source can be expressed in terms of the wavenumber spectrum of its vibrational velocity,

$$\hat{p}(x, y, z) = \frac{\rho c}{4\pi^2} \int_{-\infty}^{\infty} \int_{-\infty}^{\infty} \frac{k}{k_z} W(k_x, k_y)e^{-j(k_x x + k_y y + k_z z)} dk_x dk_y. \tag{9.219}$$

As pointed out in Section 9.5.1 Rayleigh's integral expresses the sound pressure as a convolution in the spatial domain. Equation (9.219) expresses the sound pressure as multiplication in the wavenumber domain, in agreement with the general rule that convolution in one domain corresponds to multiplication in the transformed domain.

The sound intensity can be determined in the usual manner from the sound pressure and the particle velocity. The radiated sound power can also be expressed in terms of wavenumber spectra. From Parseval's formula it follows that

$$\begin{aligned} P_a &= \frac{1}{2}\text{Re}\left\{ \int_{-\infty}^{\infty} \int_{-\infty}^{\infty} p(x, y, 0)U^*(x, y)dxdy \right\} \\ &= \frac{1}{8\pi^2}\text{Re}\left\{ \int_{-\infty}^{\infty} \int_{-\infty}^{\infty} P(k_x, k_y, 0)W^*(k_x, k_y)dk_x dk_y \right\}, \end{aligned} \tag{9.220}$$

which, with Equation (9.218), becomes

$$P_a = \frac{\rho c}{8\pi^2}\text{Re}\left\{ \int_{-\infty}^{\infty} \int_{-\infty}^{\infty} |W(k_x, k_y)|^2 \frac{k}{k_z} dk_x dk_y \right\}. \tag{9.221}$$

The z-component of the wavenumber is purely imaginary outside *the radiation circle*, that is, when

$$k_x^2 + k_y^2 > k^2, \tag{9.222}$$

and therefore the expression is reduced to

$$P_a = \frac{\rho c}{8\pi^2} \iint_{k_x^2 + k_y^2 < k^2} \frac{|W(k_x, k_y)|^2}{\sqrt{1 - (k_x/k)^2 - (k_y/k)^2}} dk_x dk_y. \tag{9.223}$$

9.5.3   Fundamentals of Near Field Acoustic Holography

Acoustic holography is an experimental technique that makes it possible to reconstruct three-dimensional sound fields from measurements on a two-dimensional surface.

The name has obviously been coined from the well-known optical technique. A particularly useful variety of acoustic holography for studying and diagnosing noise radiation problems is known as *near field acoustic holography* (often abbreviated NAH) [14]; in this method the measurement takes place on a surface fairly near the source under study (as the name implies), which makes it possible to capture evanescent waves radiated by the source. This implies a significant enhancement of the spatial resolution, as opposed to other sound visualisation techniques such as far field holography and beamforming, where the spatial resolution cannot become finer than the wavelength in the medium.

In the most fundamental version of planar near field acoustic holography one measures the sound pressure in a plane near the source under test (the aperture), calculates the corresponding wavenumber spectrum, and then reconstructs the sound field, that is, determines the sound pressure, the particle velocity and the sound intensity in other planes making use of Equations (9.215) and (9.216). This is useful in studying sound sources at low frequencies, structural vibrations below coincidence, etc. However, a number of limitations follow from the wavenumber approach: in practice the measurement area must obviously be finite whereas it is assumed to be infinite; the sound field is not measured continuously, but sampled only at a finite number of discrete positions which means that aliasing can occur (cf. Section B.7); and the two-dimensional spatial Fourier transform is approximated by a two-dimensional discrete Fourier transform which means that leakage can occur (cf. Section B.7). To reduce the influence of the finite aperture and the spatial sampling, zero padding[10] and a tapered spatial window is usually applied before the discrete spatial Fourier transform is calculated. The spatial sampling must satisfy the sampling theorem, and therefore one cannot sample *very* close to the source where there may be high spatial frequency components, resulting in aliasing. On the other hand one should not sample too far away, because then the evanescent modes will be buried in noise and cannot be reconstructed, let alone be used for backward prediction. Moreover, whereas forward holographic prediction is numerically stable, backward prediction, which involves predicting the sound field closer to the source than the measurement plane, is an unstable ill-posed inverse problem that requires regularisation because of the fact that the evanescent waves are amplified exponentially with the distance. The standard regularisation technique involves spatial low-pass filtering (multiplying with a window in the wavenumber domain).

Since the mid-1980s numerous varieties of the fundamental version of near field acoustic holography have been developed in order to overcome some of these limitations. Examples of such techniques include 'the equivalent source method', which involves adjusting the strengths of a collection of monopoles placed behind the surface of the source so as to obtain a sound field that agrees with the measurement in the measurement plane [15]; and 'statistically optimised near field acoustic holography', which is based on an expansion in elementary waves [16]; both methods altogether avoid wavenumber transforms. Another method, near field acoustic holography with the microphones mounted on a transparent or solid spherical surface, completely avoids the finite aperture problem owing to the perfect geometry of a sphere [17, 18].

The use of particle velocity transducers instead of pressure microphones (or combinations of pressure and particle velocity transducers) [19, 20, 21] has been attempted, and

---

[10] 'Zero padding' means extending the measured data with zeros. This is an efficient way of interpolating the wavenumber spectra.

it has been found that it is generally advantageous to use particle velocity transducers; the finite aperture problem is reduced since the normal component of the particle velocity dies out faster with the distance than the sound pressure does, and it is 'more sound' to predict the sound pressure from the particle velocity than the opposite, predicting the particle velocity from the particle velocity, since the latter involves spatial differentiation (which amplifies high spatial frequencies) whereas the former involves spatial integration. Moreover, it can be shown that combining the two kinds of transducers makes it possible to distinguish between sound field components from both sides of a plane array because of the fact that the particle velocity is a vector [20].

9.6 The Kirchhoff-Helmholtz Integral Equation

The Kirchhoff-Helmholtz integral equation follows from the wave equation and expresses the sound pressure in terms of an integral of the very same quantity, the sound pressure. Thus at first glance it appears to be much more complicated than the wave equation. However, it is a powerful tool that gives much insight. (It is also the background for an important numerical method of solving sound field problems, the boundary element method.) One can derive it as follows from the wave equation. Here we are concerned only with sound radiation and scattering under free-field conditions, but a similar relation can be derived for more general problems including internal problems. See, e.g., Williams [22] for a more thorough treatment.

Say we have a vibrating body in free space with the surface S, generating the sound pressure $\hat{p}(\mathbf{r})$, and let $G(\mathbf{r}, \mathbf{r}_0)$ be an arbitrary Green's function, for example, the free-space Green's function. It now follows that

$$
\begin{aligned}
G(\mathbf{r},\mathbf{r}_0)(\nabla^2 + k^2)\hat{p}(\mathbf{r}) &- \hat{p}(\mathbf{r})(\nabla^2 + k^2)G(\mathbf{r}, \mathbf{r}_0) \\
&= G(\mathbf{r}, \mathbf{r}_0)\nabla^2 \hat{p}(\mathbf{r}) - \hat{p}(\mathbf{r})\nabla^2 G(\mathbf{r}, \mathbf{r}_0) + G(\mathbf{r}, \mathbf{r}_0)k^2 \hat{p}(\mathbf{r}) - \hat{p}(\mathbf{r})k^2 G(\mathbf{r}, \mathbf{r}_0) \quad (9.224)\\
&= \nabla \cdot (G(\mathbf{r}, \mathbf{r}_0)\nabla \hat{p}(\mathbf{r}) - \hat{p}(\mathbf{r})\nabla G(\mathbf{r}, \mathbf{r}_0)).
\end{aligned}
$$

(This is known as Green's theorem.) If we integrate this expression over a large volume V that encloses but does not include the body of the source, the first term on the left-hand side vanishes because $\hat{p}$ satisfies the Helmholtz equation, whereas the second term on the left-hand side integrates to $\hat{p}(\mathbf{r}_0)$ since G satisfies the inhomogeneous Helmholtz equation with a delta function on the right-hand side (see Example 9.1). Gauss's theorem turns the right-hand term of Equation (9.224) to an integral of the normal component of $G\nabla\hat{p} - \hat{p}\nabla G$ over the surface of the volume, that is, the interior surface S and the outer surface of the volume – see Figure 9.27. One may argue that the latter surface integral must vanish in the limit of a large enough volume since both the sound pressure generated by the source and the Green's function satisfy the Sommerfeld radiation condition; thus the contribution from the integral over the outer surface must be zero. It follows that

$$
\begin{aligned}
\hat{p}(\mathbf{r}_0) &= \int_V \nabla \cdot (G(\mathbf{r}, \mathbf{r}_0)\nabla \hat{p}(\mathbf{r}) - \hat{p}(\mathbf{r})\nabla G(\mathbf{r}, \mathbf{r}_0))dV \\
&= \int_S (G(\mathbf{r}, \mathbf{r}_0)\nabla \hat{p}(\mathbf{r}) - \hat{p}(\mathbf{r})\nabla G(\mathbf{r}, \mathbf{r}_0)) \cdot \mathbf{n}\, dS \quad (9.225)\\
&= -\int_S \left(j\omega\rho \hat{u}_n(\mathbf{r})G(\mathbf{r}, \mathbf{r}_0) + \hat{p}(\mathbf{r})\frac{\partial G(\mathbf{r}, \mathbf{r}_0)}{\partial n} \right) dS
\end{aligned}
$$

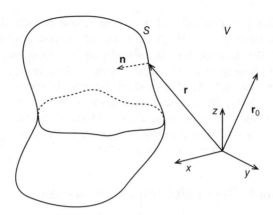

Figure 9.27 A body with the surface S in an infinite volume V

if $\mathbf{r}_0$ is exterior to the surface of the source, S. Note, however, that one cannot specify both the normal velocity and the sound pressure independently on S; if, say, the velocity of the source is given then the sound pressure on the surface follows. In other words, Equation (9.225) is not a solution since the unknown sound pressure appears on both sides of the equation; it is an integral equation that must be solved.

Comparing Equations (9.2) and (9.12) leads to the conclusion that one may interpret the first term on the right-hand side of Equation (9.225) as the contribution due to a collection of monopoles, each with the strength $\hat{u}_n \mathrm{d}S$; and the second term as due to a collection of dipoles, each with the strength $(\hat{p}/(\rho c k))\mathrm{d}S$. With a given normal velocity on the surface of the source the second term takes account of the fact that the sound field generated by each monopole is scattered by the body of the source.

In the case of a scattering problem one simply adds the undisturbed incident sound field to the right-hand side of Equation (9.225):

$$\hat{p}(\mathbf{r}_0) = -\int_S \left(j\omega\rho\hat{u}_n(\mathbf{r})G(\mathbf{r},\mathbf{r}_0) + \hat{p}(\mathbf{r})\frac{\partial G(\mathbf{r},\mathbf{r}_0)}{\partial n} \right) \mathrm{d}S + \hat{p}_{\mathrm{inc}}(\mathbf{r}_0). \qquad (9.226)$$

If the scattering body is rigid the first term of the integrand vanishes because the normal component of the gradient of the pressure must be zero. Note, however, that, as in Sections 9.3.2 and 9.4.2, the sound pressure is the total pressure; the gradient of the incident sound pressure must be counterbalanced by the gradient of the scattered pressure.

Equation (9.225) can be simplified considerably if it is possible to replace the free-space Green's function with a Green's function such that G or $\partial G/\partial n$ and thus the first or second term of the integrand vanishes on the integration surface. We will make use of this in what follows. Rayleigh's second integral, which expresses the sound pressure in terms of the normal velocity on an infinite plane (Equation (9.191)), may be regarded as a special case of the Kirchhoff-Helmholtz integral equation. In this case it is convenient to use a Green's function that takes account of the rigid plane, G' instead of the free-space Green's function G. This involves an image source placed symmetrically on the other side of the rigid surface,

$$G'(\mathbf{r},\mathbf{r}_0) = G(\mathbf{r},\mathbf{r}_0) + G(\mathbf{r},\mathbf{r}_0'), \qquad (9.227)$$

where $\mathbf{r}_0'$ is the position of the image source. This leads to the conclusion that G' equals twice the free-field Greens function on the rigid surface whereas the normal component of the derivative, $\partial G'/\partial n$, will be zero. It follows that Equation (9.225) becomes

$$\hat{p}(\mathbf{r}_0) = \int_S j\omega\rho\hat{u}_n(\mathbf{r})G'(\mathbf{r},\mathbf{r}_0)\mathrm{d}S = 2\int_S j\omega\rho\hat{u}_n(\mathbf{r})G(\mathbf{r},\mathbf{r}_0)\mathrm{d}S$$
$$= \int_S j\omega\rho\hat{u}_n(\mathbf{r})\frac{e^{-jkR}}{2\pi R}\mathrm{d}S, \tag{9.228}$$

where $R = |\mathbf{r} - \mathbf{r}_0|$ and the direction of the normal has been reversed. Equation (9.228) is recognised as the Rayleigh integral.

Rayleigh's *first* integral is another special case. It describes the sound field that can be predicted if we have an infinite plane surface with a prescribed sound pressure. For this case we will use a Green's function that takes account of an infinite pressure-release surface, which involves a negative image source placed symmetrically,

$$G''(\mathbf{r},\mathbf{r}_0) = G(\mathbf{r},\mathbf{r}_0) - G(\mathbf{r},\mathbf{r}_0'), \tag{9.229}$$

where (again) $\mathbf{r}_0'$ is the position of the image source. In this case G'' vanishes on the surface whereas the normal component of its gradient, $\partial G''/\partial n$, equals twice the gradient of the free-space Green's function. The result is

$$\hat{p}(\mathbf{r}_0) = \int_S \hat{p}(\mathbf{r})\frac{\partial G''(\mathbf{r},\mathbf{r}_0)}{\partial n}\mathrm{d}S = \int_S 2\hat{p}(\mathbf{r})\frac{\partial G(\mathbf{r},\mathbf{r}_0)}{\partial n}\mathrm{d}S$$
$$= \int_S \hat{p}(\mathbf{r})\frac{jke^{-jkR}}{2\pi R}\left(1+\frac{1}{jkR}\right)\cos\theta\ \mathrm{d}S, \tag{9.230}$$

where θ is the angle between the vector $\mathbf{r} - \mathbf{r}_0$ and the normal vector to the plane. Equation (9.230) expresses the sound pressure in one plane in terms of the sound pressure in another plane, just as Equation (9.215) does.

References

[1] A.D. Pierce *Acoustics: An Introduction to Its Physical Principles and Applications*. 2nd edition, Acoustical Society of America, New York (1989).

[2] P.A. Nelson and S.J. Elliott: *Active Control of Sound*. Academic Press, London (1992).

[3] G. Elko and J. Meyer, Microphone arrays. Chapter 50 in *Springer Handbook of Speech Processing*, ed. J. Benesty, M. Sondhi, and Y. Huang. Springer-Verlag, Berlin (2008).

[4] E. Tiana-Roig, F. Jacobsen, and E. Fernandez Grande: Beamforming with a circular microphone array for localization of environmental noise sources. *Journal of the Acoustical Society of America* **128**, 3535–3542 (2010).

[5] J. Meyer: Beamforming for a circular microphone array mounted on spherically shaped objects. *Journal of the Acoustical Society of America* **109**, 185–193 (2001).

[6] M. Park and B. Rafaely: Sound-field analysis by plane-wave decomposition using spherical microphone array. *Journal of the Acoustical Society of America* **118**, 3094–3103 (2005).

[7] H. Teutsch and W. Kellermann: Acoustic source detection and localization based on wavefield decomposition using circular microphone arrays. *Journal of the Acoustical Society of America* **120**, 2724–2736 (2006).

[8] G.A. Daigle, M.R. Stinson, and J.G. Ryan: Beamforming with aircoupled surface waves around a sphere and circular cylinder. *Journal of the Acoustical Society of America* **117**, 3373–3376 (2005).

[9] K. Ehrenfried, L. Koop, A comparison of iterative deconvolution algorithms for the mapping of acoustic sources, *AIAA Journal* **45**, 1584–1595 (2007).

[10] F. Jacobsen: A note on instantaneous and time-averaged active and reactive sound intensity. *Journal of Sound and Vibration* **147**, 489–496, 1991.

[11] M.A. Gerzon: Periphony width-height sound reproduction. *Journal of the Audio Engineering Society* **21**, 2–10 (1973).

[12] J. Ahrens and S. Spors: An analytical approach to sound field reproduction using circular and spherical loudspeaker distributions. *Acta Acustica united with Acustica* **94**, 988–999 (2008).

[13] J. Daniel, R. Nicol and S. Moreau: Further investigations of higher-order ambisonics and wavefield synthesis for holophonic sound imaging. *114th Convention of the Audio Engineering Society*, paper 5788, Amsterdam (2003).

[14] J.D. Maynard, E.G. Williams and Y. Lee: Nearfield acoustic holography. I. Theory of generalized holography and the development of NAH. *Journal of the Acoustical Society of America* **78**, 1395–1413 (1985).

[15] G.H. Koopmann, L. Song, and J.B. Fahnline: A method for computing acoustic fields based on the principle of wave superposition. *Journal of the Acoustical Society of America* **86**, 2433–2438 (1989).

[16] J. Hald: Basic theory and properties of statistically optimized near-field acoustical Holography. *Journal of the Acoustical Society of America* **125**, 2105–2120 (2009).

[17] E.G. Williams, N. Valdivia, and P. C. Herdic: Volumetric acoustic vector intensity imager. *Journal of the Acoustical Society of America* **120**. 1887–1897 (2006).

[18] E.G. Williams and K. Takashima: Vector intensity reconstructions in a volume surrounding a rigid spherical microphone array. *Journal of the Acoustical Society of America* **127**, 773–783 (2010).

[19] F. Jacobsen and Y. Liu: Near field acoustic holography with particle velocity transducers. *Journal of the Acoustical Society of America* **118**, 3139–3144 (2005).

[20] F. Jacobsen and V. Jaud: Statistically optimized near field acoustic holography using an array of pressure-velocity probes. *Journal of the Acoustical Society of America* **121**, 1550–1558 (2007).

[21] E. Fernandez-Grande and F. Jacobsen, Sound field separation with a double layer velocity transducer array. *Journal of the Acoustical Society of America* **130**, 5–8 (2011).

[22] E.G. Williams: *Fourier Acoustics: Sound Radiation and Nearfield Acoustical Holography*, Academic Press, London (1999).

Appendix A

Complex Representation of Harmonic Functions of Time

In a harmonic sound field the sound pressure at any point is a function of the type $\cos(\omega t + \varphi)$. It is common practice to use *complex notation* in such cases. This is a symbolic method that makes use of the fact that complex exponentials give a more condensed notation than trigonometric functions because of the more complicated multiplication theorems of sines and cosines.

We recall that a complex number A can be written either in terms of its real and imaginary part or in terms of its magnitude (also called absolute value or modulus) and phase angle,

$$A = A_r + jA_i = |A|e^{j\varphi_A}, \tag{A.1}$$

where

$$j = \sqrt{-1} \tag{A.2}$$

is the imaginary unit, and

$$A_r = \text{Re}\{A\} = |A|\cos\varphi_A, \quad A_i = \text{Im}\{A\} = |A|\sin\varphi_A, \tag{A.3, A.4}$$

$$|A| = \sqrt{A_r^2 + A_i^2} \tag{A.5}$$

(see Figure A.1). The complex conjugate of A is

$$A^* = A_r - jA_i = |A|e^{-j\varphi_A}; \tag{A.6}$$

therefore the magnitude can also be written

$$|A| = \sqrt{A \cdot A^*}. \tag{A.7}$$

Multiplication and division of two complex numbers are most conveniently carried out if they are given in terms of magnitudes and phase angles,

$$AB = |A||B|e^{j(\varphi_A + \varphi_B)}, \quad A/B = \frac{|A|}{|B|}e^{j(\varphi_A - \varphi_B)}. \tag{A.8, A.9}$$

Fundamentals of General Linear Acoustics, First Edition. Finn Jacobsen and Peter Møller Juhl.
© 2013 John Wiley & Sons, Ltd. Published 2013 by John Wiley & Sons, Ltd.

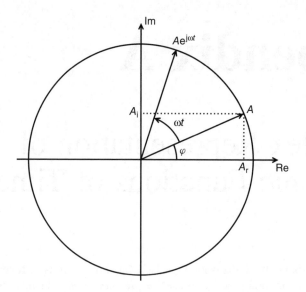

Figure A.1   Complex representation of a harmonic signal

Complex representation of harmonic signals makes use of the fact that

$$e^{jx} = \cos x + j \sin x \tag{A.10}$$

(Euler's equation) or, conversely,

$$\cos x = \frac{1}{2}(e^{jx} + e^{-jx}), \quad \sin x = -j\frac{1}{2}(e^{jx} - e^{-jx}). \tag{A.11a, A.11b}$$

In a harmonic sound field the sound pressure at a given position can be written

$$\hat{p} = Ae^{j\omega t}, \tag{A.12}$$

where A is the *complex amplitude* of the sound pressure. The real, physical sound pressure is of course a real function of the time,

$$p = \text{Re}\{\hat{p}\} = \text{Re}\{|A| e^{j(\omega t + \phi_A)}\} = |A| \cos(\omega t + \phi_A), \tag{A.13}$$

which is seen to be an expression of the form $\cos(\omega t + \varphi)$. The magnitude of the complex quantity $|A|$ is called the *amplitude* of the pressure, and φ_A is its phase. It can be concluded that complex notation implies the mathematical trick of adding another solution, an expression of the form $\sin(\omega t + \varphi)$, multiplied by a constant, the imaginary unit j. This trick relies on linear superposition.

The mathematical convenience of the complex representation of harmonic signals can be illustrated by an example. A sum of two harmonic signals of the same frequency, $A_1 e^{j\omega t}$ and $A_2 e^{j\omega t}$, is yet another harmonic signal with an amplitude of $|A_1 + A_2|$

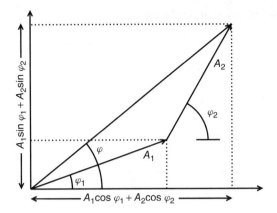

Figure A.2   Two simple harmonic signals with identical frequencies

(see Figure A.2). Evidently, this can also be derived without complex notation,

$$
\begin{aligned}
p &= |A_1| \cos(\omega t + \varphi_1) + |A_2| \cos(\omega t + \varphi_2) \\
&= (|A_1| \cos \varphi_1 + |A_2| \cos \varphi_2) \cos \omega t - (|A_1| \sin \varphi_1 + |A_2| \sin \varphi_2) \sin \omega t \\
&= \left[(|A_1| \cos \varphi_1 + |A_2| \cos \varphi_2)^2 + (|A_1| \sin \varphi_1 + |A_2| \sin \varphi_2)^2 \right]^{1/2} \cos(\omega t + \varphi),
\end{aligned}
$$

$$(A.14)$$

where

$$
\varphi = \arctan \frac{|A_1| \sin \varphi_1 + |A_2| \sin \varphi_2}{|A_1| \cos \varphi_1 + |A_2| \cos \varphi_2},
\tag{A.15}
$$

but the expedience and convenience of the complex method seems indisputable.
    Since

$$
\frac{d}{dt} e^{j\omega t} = j\omega e^{j\omega t},
\tag{A.16}
$$

it follows that differentiation with respect to time corresponds to multiplication by a factor of $j\omega$. Conversely, integration with respect to time corresponds to division with $j\omega$. If, for example, the vibrational velocity of a surface is, in complex representation,

$$
\hat{v} = B e^{j\omega t} = |B| e^{j(\omega t + \varphi_B)},
\tag{A.17}
$$

which means that the real, physical velocity is

$$
v = \mathrm{Re}\{\hat{v}\} = |B| \cos(\omega t + \phi_B),
\tag{A.18}
$$

then the acceleration is written

$$
\hat{a} = j\omega \hat{v},
\tag{A.19}
$$

which means that the physical acceleration is

$$a = \mathrm{Re}\{\hat{a}\} = \mathrm{Re}\{j\omega B e^{j\omega t}\} = -\omega |B| \sin(\omega t + \varphi_B), \tag{A.20}$$

and this is seen to agree with the fact that

$$\frac{d}{dt}\cos(\omega t + \varphi_B) = -\omega \sin(\omega t + \varphi_B). \tag{A.21}$$

In a similar manner we find the displacement,

$$\hat{\xi} = \frac{\hat{v}}{j\omega}, \tag{A.22}$$

which means that

$$\xi = \mathrm{Re}\{\hat{\xi}\} = \mathrm{Re}\left\{\frac{1}{j\omega}B e^{j\omega t}\right\} = \frac{1}{\omega}|B|\sin(\omega t + \varphi_B), \tag{A.23}$$

in agreement with the fact that

$$\frac{d}{dt}\left(\frac{1}{\omega}\sin\left(\omega t + \varphi_B\right)\right) = \cos(\omega t + \varphi_B). \tag{A.24}$$

Acoustic second-order quantities involve time averages of squared harmonic signals and, more generally, products of harmonic signals. Such quantities are dealt with in a special way, as follows. Expressed in terms of the complex pressure amplitude $\hat{p}$, the mean square pressure becomes

$$\langle p^2(t)\rangle_t = p_{\mathrm{rms}}^2 = |\hat{p}|^2/2, \tag{A.25}$$

in agreement with the fact that the average value of a squared cosine is $1/2$. Note that it is the squared *magnitude* of $\hat{p}$ that enters into the expression, not the square of $\hat{p}$, which in general would be a complex number proportional to $e^{2j\omega t}$.

The time average of a product is given by the following expression

$$\langle x(t)y(t)\rangle_t = \frac{1}{2}\mathrm{Re}\{\hat{x}\hat{y}^*\} = \frac{1}{2}\mathrm{Re}\{\hat{x}^*\hat{y}\}. \tag{A.26}$$

This can be seen as follows,

$$\frac{1}{2}\mathrm{Re}\left\{\hat{x}\hat{y}^*\right\} = \frac{1}{2}\mathrm{Re}\left\{|\hat{x}|\,e^{j(\omega t + \phi_x)}|\hat{y}|e^{-j(\omega t + \phi_y)}\right\} = \frac{1}{2}|\hat{x}||\hat{y}|\cos(\phi_x - \phi_y), \tag{A.27}$$

which is seen to agree with

$$\langle x(t)y(t)\rangle_t = \langle|\hat{x}|\cos(\omega t + \phi_x)|\hat{y}|\cos(\omega t + \phi_y)\rangle_t = \frac{1}{2}|\hat{x}||\hat{y}|\cos(\phi_x - \phi_y). \tag{A.28}$$

Note that either $\hat{x}$ or $\hat{y}$ must be conjugated.

Appendix B

Signal Analysis and Processing

B.1 Introduction

Experiments play an important part in acoustics. In acoustic experiments use is very often made of concepts from signal analysis and measurement techniques based on digital signal processing. For example, the acoustic properties of a combination of tubes may be determined by driving it with source with a known volume velocity and impedance and measuring the resulting sound pressure at various positions; and the dynamic properties of a structure may be determined by exciting it with a known time-varying force and measuring the vibrational velocity at various positions. In some branches of acoustics, room acoustics in particular, the analysis is usually performed in the time domain and the information is deduced from the impulse response; in other fields the analysis is usually carried out in the frequency domain. However, use of equipment such as, e.g., a multi-channel FFT analyser presupposes a certain knowledge of signal analysis. The purpose of this Appendix is to give a brief and elementary introduction to some of the most fundamental concepts and methods from this discipline with particular regard to experimental determination of the properties of acoustic systems.

B.2 Classification of Signals

In the following a *signal* will mean a continuous function of time.[1] The signal represents a time-varying physical quantity, for instance the sound pressure at a certain position in a sound field or the vibrational velocity of a structure excited by a time-varying force; therefore only *real* signals will be considered here. The signals that will be dealt with in this note can be divided into three types (see Figure B.1):

a) transient signals,
b) periodic signals,
c) stationary random signals.

---

[1] Such signals are often referred to as *analogue* (as opposed to digital) signals.

Fundamentals of General Linear Acoustics, First Edition. Finn Jacobsen and Peter Møller Juhl.
© 2013 John Wiley & Sons, Ltd. Published 2013 by John Wiley & Sons, Ltd.

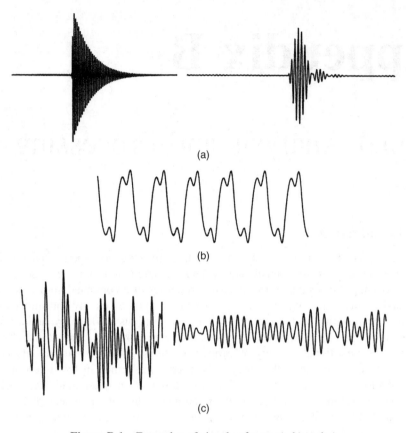

Figure B.1   Examples of signals of type a), b) and c)

The division into different groups is useful because the three types of signals require different mathematical descriptions and must be analysed with different types of analysis.

B.3   Transient Signals

A transient signal is a signal with finite total energy. From a practical point of view such signals are of finite duration (although the duration of an impulse that decays exponentially, say, is a matter of definition). The energy[2] of the signal $x(t)$ is

$$E_x = \int_{-\infty}^{\infty} x^2(t) \ \mathrm{d}t. \tag{B.1}$$

Transient signals can be described in the frequency domain in terms of their *Fourier spectra*.

---

[2] It is customary to use the term 'energy' in the sense of the integral of the square of the signal, without regard to its particular units.

B.3.1   The Fourier Transform

The spectrum (or, strictly speaking, the spectral density or the Fourier transform) of a transient signal $x(t)$ is a complex function of the frequency defined through the integral

$$X(\omega) = \int_{-\infty}^{\infty} e^{-j\omega t}\,dt, \tag{B.2}$$

where $\omega = 2\pi f$ is the angular frequency. There is a unique relation between a signal and its spectrum. The inverse Fourier transform brings us back from the frequency domain to the time domain:

$$x(t) = \frac{1}{2\pi} \int_{-\infty}^{\infty} X(\omega)e^{j\omega t}\,d\omega. \tag{B.3}$$

In this Appendix we shall use the notation

$$x(t) \quad \leftrightarrow \quad X(\omega) \tag{B.4}$$

for a Fourier transform pair.

Note that $X(\omega)$ is a complex-valued function of both positive and negative frequencies,

$$X(\omega) = \mathrm{Re}\{X(\omega)\} + j\mathrm{Im}\{X(\omega)\} = |X(\omega)|e^{j\varphi_x(\omega)}, \tag{B.5}$$

where the function $|X(\omega)|$ is the amplitude spectrum and $\varphi_x(\omega)$ is the phase spectrum. However, since $x(t)$ is real it can be seen from Equation (B.2) that $\mathrm{Re}\{X(\omega)\}$ and $|X(\omega)|$ are even functions of the frequency whereas $\mathrm{Im}\{X(\omega)\}$ and $\varphi_x(\omega)$ are odd functions (see Figure B.2); therefore

$$X(-\omega) = X^*(\omega), \tag{B.6}$$

which shows that there is no additional information about the signal at the negative frequencies. However, *two-sided* spectral density functions are mathematically convenient, as will become apparent.

If $x(t)$ is an even function then $X(\omega)$ is purely real; if $x(t)$ is an odd function then $X(\omega)$ is purely imaginary.

Example B.1 An exponential decay and its Fourier transform: If $x(t)$ is an exponential decay,

$$x(t) = \begin{cases} 0 & \text{if } t < 0 \\ ae^{-\beta t} & \text{elsewhere} \end{cases}$$

(where β is real and positive), then

$$X(\omega) = a \int_{-\infty}^{\infty} e^{-\beta t} e^{-j\omega t}\,dt = \frac{a}{\beta + j\omega}.$$

This Fourier transform pair is shown in Figure B.2.

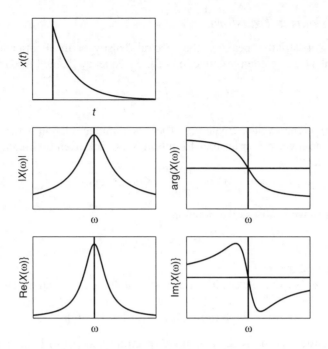

Figure B.2 An exponentially decaying pulse and its Fourier transform, shown as amplitude and phase or real and imaginary part

The following simple theorems are useful. They can all be derived from Equations (B.2) and (B.3) (see, e.g., Reference [1]), but are given here without proof.

It is evident from Equations (B.2) and (B.3) that there is a strong similarity between the Fourier transform and the inverse Fourier transform. If

$$x(t) \quad \leftrightarrow \quad X(\omega) \tag{B.7}$$

then

$$X(t) \quad \leftrightarrow \quad 2\pi x(-\omega). \tag{B.8}$$

This is known as the *symmetry theorem*.

Example B.2 The symmetry theorem: The Fourier transform of the rectangular pulse,

$$x(t) = \begin{cases} a & \text{if } |t| < \dfrac{T}{2} \\ 0 & \text{elsewhere,} \end{cases}$$

is

$$X(\omega) = a \int_{-T/2}^{T/2} e^{-j\omega t} dt = aT \frac{\sin \dfrac{\omega T}{2}}{\dfrac{\omega T}{2}}$$

Conversely, the transient signal

$$x(t) = a\frac{\sin \beta t}{\beta t}$$

corresponds to the spectrum

$$X(\omega) = \begin{cases} \dfrac{\pi a}{\beta} & \text{if } |\omega| < \beta \\ 0 & \text{elsewhere,} \end{cases}$$

see Figure B.3.

For any real value of a,

$$x(at) \quad \leftrightarrow \quad \frac{1}{|a|}X\left(\frac{\omega}{a}\right) \tag{B.9}$$

(see Figure B.4). Obviously, compression in the time domain corresponds to expansion in the frequency domain (and vice versa). Note also the effect of time reversal:

$$x(-t) \leftrightarrow X(-\omega) = X^*(\omega). \tag{B.10}$$

This is the *time scaling theorem*.

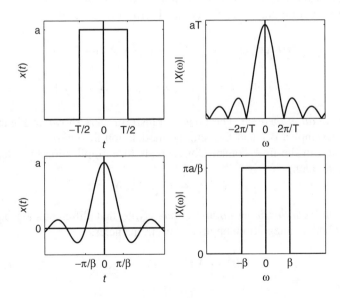

Figure B.3 Illustration of the symmetry between the Fourier transform and the inverse Fourier transform

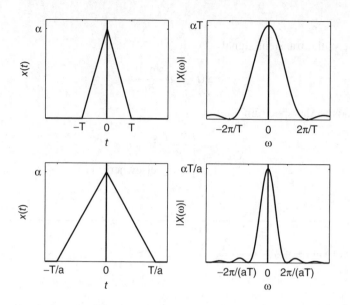

Figure B.4   The influence of time scaling

A delay of the signal has no influence on the amplitude spectrum, but the phase is modified:

$$x(t - t_o) \quad \leftrightarrow \quad X(\omega)e^{-j\omega t_0}. \tag{B.11}$$

Compare Figures B.2 and B.5. This is the *time shifting theorem*.
Multiplication with a pure-tone signal has the following effect:

$$x(t)\cos \omega_0 t \leftrightarrow \frac{1}{2}(X(\omega - \omega_0) + X(\omega + \omega_0)), \tag{B.12a}$$

$$x(t)\sin \omega_0 t \leftrightarrow \frac{1}{2j}(X(\omega - \omega_0) - X(\omega + \omega_0)). \tag{B.12b}$$

This is the *modulation theorem*. The Fourier transform pairs shown in Figure B.6 can be derived from Examples B.1 and B.2 and the modulation theorem.
Differentiation in the time domain corresponds to multiplication by a factor of $j\omega$ in the frequency domain:

$$x^{(n)}(t) \leftrightarrow (j\omega)^n X(\omega). \tag{B.13}$$

This is the *differentiation/integration theorem*. Note that differentiation amplifies high frequencies, whereas integration amplifies low frequencies.
If

$$x(t) \leftrightarrow X(\omega),\; y(t) \leftrightarrow Y(\omega), \tag{B.14a, B.14b}$$

then

$$x(t) * y(t) = \int_{-\infty}^{\infty} x(u)y(t - u)du \leftrightarrow X(\omega)Y(\omega). \tag{B.15}$$

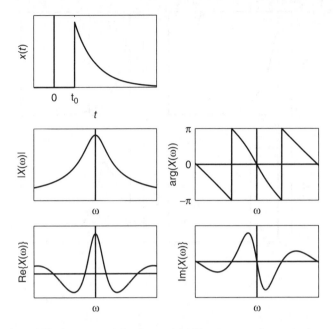

Figure B.5   A time-shifted exponential pulse and its Fourier transform, shown as amplitude and phase or real and imaginary part

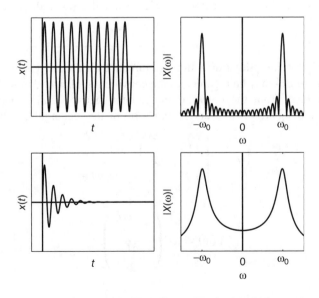

Figure B.6   The effect of modulation

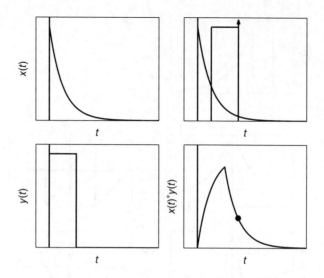

Figure B.7 The convolution of two functions

Conversely,

$$x(t)y(t) \leftrightarrow \frac{1}{2\pi} X(\omega) * Y(\omega). \tag{B.16}$$

This is the *convolution theorem*, according to which multiplication in one domain corresponds to convolution in the other domain. Figure B.7 illustrates the convolution of two functions of time.

Example B.3 A triangular pulse and its Fourier transform: The triangular pulse shown in Figure B.4 may be regarded as the convolution of two identical rectangular pulses. It follows from the convolution theorem that the corresponding Fourier transform is the square of the Fourier transform of the rectangular pulse: if

$$x(t) = \begin{cases} a\left(1 - \dfrac{|t|}{T}\right) & \text{if } |t| < T \\ 0 & \text{elsewhere,} \end{cases}$$

then

$$X(\omega) = aT\left(\frac{\sin \dfrac{\omega T}{2}}{\dfrac{\omega T}{2}}\right)^2.$$

From Equations (B.10) and (B.15) it follows that

$$x(t) * x(-t) \leftrightarrow |X(\omega)|^2, \tag{B.17}$$

that is, the convolution of a transient signal with a time-reversed version of the signal is identical with the inverse Fourier transform of the square of the amplitude spectrum:

$$\int_{-\infty}^{\infty} x(u)x(u-t)\,du = \frac{1}{2\pi}\int_{-\infty}^{\infty} X(\omega)^2 e^{j\omega t}\,d\omega. \tag{B.18}$$

With $t = 0$, Equation (B.18) simplifies to

$$E_x = \int_{-\infty}^{\infty} x^2 dt = \frac{1}{2\pi}\int_{-\infty}^{\infty} |X(\omega)|^2 d\omega, \tag{B.19}$$

which shows that the energy of a transient signal can be obtained either by integrating the squared signal over all time or by integrating the squared modulus of the spectrum over all frequencies (Parseval's formula). Accordingly, $|X(\omega)|^2$ is called the *energy spectrum* (or, more precisely, the energy spectral density).

B.3.2 Time Windows

In analysing a signal of long duration, a stationary signal, for example, or a very long impulse, one is often faced with the problem of estimating the properties of the signal from a short sample. Cutting off a segment of a signal or dividing a signal into a number of segments is called 'windowing'. The effect of such a process can be analysed by means of Equation (B.16) since, apparently, the signal is multiplied by a weighting function, a 'time window'. Since multiplication in the time domain corresponds to convolution in the frequency domain, it follows that the spectrum of the signal is convolved with the spectrum of the time window. Obviously, the effect of this smoothing depends strongly on the particulars of the spectrum of the signal: spectra with sharp peaks or troughs are particularly likely to be distorted. However, it is generally advantageous to use a time window with a concentrated spectrum that decays rapidly.

Figure B.8 shows three different time windows and their spectra. The rectangular window, which simply chops off a part of the signal, has the disadvantage that its spectrum decays relatively slowly. It is apparent that the Hanning window, that is, the function

$$v(t) = \begin{cases} \sin^2\dfrac{t\pi}{T} & \text{if } 0 \le t \le T \\ 0 & \text{elsewhere,} \end{cases} \tag{B.20}$$

is more favourable: the main lobe is broader, but the sidelobes decay far more rapidly. This window is in fact the most commonly used weighting function in analysis of stationary signals. See also Figures B.13 and B.14 in Section B.4.

Time windows can also be used in separating various components of a transient signal from each other ('gating'), say, separating an impulse that is due to a direct sound wave from other pulses that are due to reflected waves. In that case a tapered rectangular function is often used. If it is necessary to shorten a signal and it is known from physical reasons that the signal consists of a number of exponentially damped modes, then it may be advantageous to apply an exponential time window (see Figure B.2); this process corresponds to artificially increasing the damping of the tones, which is a relatively mild degradation of the signal.

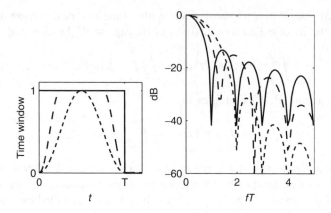

Figure B.8 Examples of commonly used time windows and their spectra. Solid line, the rectangular window; dashed line, a cosine-tapered rectangular window; dotted line, the Hanning window

B.4 Periodic Signals

A periodic signal $x(t)$ is a signal of infinite duration for which

$$x(t + nT) = x(t), \tag{B.21}$$

where n is any integer and T is the period, that is, the signal is completely characterised by its properties within an interval of the duration T. The power[3] (or mean square value) of the signal is

$$P_x = \frac{1}{T} \int_{-T/2}^{T/2} x^2(t)\,\mathrm{d}t \tag{B.22}$$

(cf. Equation (B.2)). The square root of P_x is called the rms value (root mean square).

B.4.1 Fourier Series

A periodic signal with the fundamental frequency $\omega_0 = 2\pi/T$ can be described in terms of a Fourier series, that is, it can be decomposed into harmonic frequency components corresponding to the frequencies $\omega_0, 2\omega_0, 3\omega_0, \ldots,$

$$x(t) = \sum_{n=-\infty}^{\infty} a_n \mathrm{e}^{\mathrm{j}n\omega_0 t}, \tag{B.23}$$

where the complex Fourier coefficients a_n are given by

$$a_n = \frac{1}{T} \int_{-T/2}^{T/2} x(t)\mathrm{e}^{-\mathrm{j}n\omega_0 t}\,\mathrm{d}t. \tag{B.24}$$

---

[3] With stationary signals it is customary to use the term 'power' in the sense of the average value of the square of the signal, irrespective of its units.

Note that $a_n = a^*_{-n}$. Equation (B.23) can also be written in the form

$$x(t) = a_0 + 2\sum_{n=1}^{\infty} |a_n| \cos(n\omega_0 t + \varphi_n),$$ (B.25)

where φ_n is the phase angle of a_n,

$$a_n = |a_n| e^{j\varphi_n}.$$ (B.26)

Equation (B.25) demonstrates somewhat more directly than Equation (B.23) that a periodic signal can be decomposed into sinusoidal components at discrete frequencies that are multiples of the fundamental frequency; see Figure B.9. However, Equation (B.23) is mathematically very convenient.

It can be shown that

$$P_x = \frac{1}{T} \int_{-T/2}^{T/2} x^2(t) \mathrm{d}t = \sum_{n=-\infty}^{\infty} |a_n^2|$$ (B.27)

(Parseval's formula), which means that the total power of the signal equals the sum of the powers of all the harmonic components (cf. Equation (B.19)).

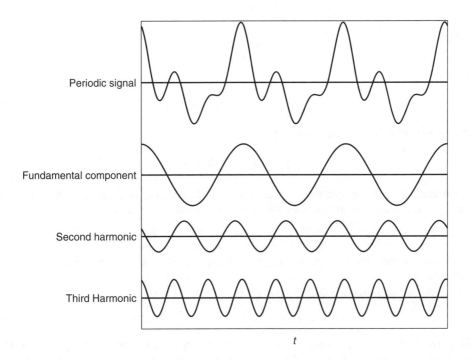

Figure B.9   A periodic signal consisting of three harmonic components.

B.4.2   The Fourier Transform of a Periodic Signal

It is not self-evident that periodic signals can be Fourier transformed, since integrals of
the form

$$X(\omega) = \int_{-\infty}^{\infty} \cos(n\omega_0 t) e^{-j\omega t} dt = \int_{-\infty}^{\infty} \cos(n\omega_0 t)(\cos \omega t - j \sin \omega t) dt \qquad (B.28)$$

have no obvious meaning. However, it is indeed possible to define the Fourier spectrum
of a periodic signal if the *delta function*, or the unit impulse, is introduced.

The delta function $\delta(\omega)$ belongs to a class of functions that are called 'distributions'
or 'generalised functions'. These functions cannot be specified in terms of their values
but must be specified in terms of their properties [2]. The property that defines the delta
function is specified by the equation

$$\int_{-\infty}^{\infty} \delta(t)x(t) dt = x(0), \qquad (B.29)$$

where $x(t)$ is an arbitrary continuous function.[4] The delta function may be regarded as
the limiting case of a pulse with unit 'area' (time integral) that becomes infinitely short
but retains its area.

From Equation (B.29) it follows that

$$\int_{-\infty}^{\infty} \delta(t - t_0)x(t) dt = x(t_0) \qquad (B.30)$$

and that

$$\delta(t) * x(t) = \int_{-\infty}^{\infty} \delta(u)x(t - u) du = x(t) \qquad (B.31)$$

and

$$\delta(t - t_0) * x(t) = x(t - t_0). \qquad (B.32)$$

The Fourier transform of the delta function is

$$\int_{-\infty}^{\infty} \delta(t) e^{-j\omega t} dt = 1, \qquad (B.33)$$

which implies that the spectrum of an infinitely short pulse is constant and extents to
infinity. Conversely, from Equation (B.8) it follows that

$$1 \leftrightarrow 2\pi\delta(\omega), \qquad (B.34)$$

which shows that the Fourier spectrum of a DC-signal is a delta function at the frequency zero.

Making use of Equation (B.12) now gives

$$\cos(\omega_0 t) \leftrightarrow \pi(\delta(\omega - \omega_0) + \delta(\omega + \omega_0)). \qquad (B.35)$$

Equation (B.35) demonstrates that the Fourier spectrum of a sinusoidal signal with the
frequency ω_0 consists of two delta functions, one at ω_0 and one at $-\omega_0$, as shown in
Figure B.10.

---

[4] Note that the delta function is not dimensionless; $\delta(\omega)dt$ is dimensionless.

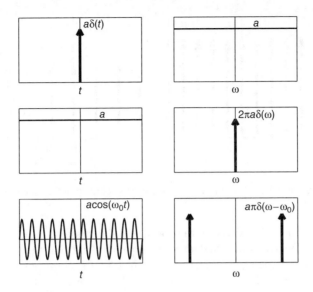

Figure B.10 Examples of Fourier transform pairs: an impulse, a DC-signal and a pure tone

Any periodic signal may be regarded as a periodically repeated transient signal, which means that it can be written in the form of a transient convolved with a 'train' of impulses:

$$x(t) = \sum_{m=-\infty}^{\infty} x_p(t - mT) = x_p(t) * \sum_{m=-\infty}^{\infty} \delta(t - mT) \qquad (B.36)$$

(cf. Equation (B.15)). It can be shown that the Fourier spectrum of such a train of impulses is a similar train of impulses in the frequency domain [1]:

$$\sum_{m=-\infty}^{\infty} \delta(t - mT) \leftrightarrow \omega_0 \sum_{n=-\infty}^{\infty} \delta(\omega - n\omega_0). \qquad (B.37)$$

Combining Equations (B.15), (B.36) and (B.37) gives the following result:

$$x(t) = x_p(t) * \sum_{m=-\infty}^{\infty} \delta(t - mT) \leftrightarrow \omega_0 X_p(\omega) \sum_{n=-\infty}^{\infty} \delta(\omega - n\omega_0)$$

$$= \omega_0 \sum_{n=-\infty}^{\infty} X_p(n\omega_0)\delta(\omega - n\omega_0) \qquad (B.38)$$

where

$$x_p(t) \leftrightarrow X_p(\omega). \qquad (B.39)$$

It can be concluded that each Fourier coefficient of the repeated transient is related to the spectral density of the transient signal $x_p(t)$, sampled at the corresponding frequency:

$$a_n = \frac{1}{T} X_p(n\omega_0). \qquad (B.40)$$

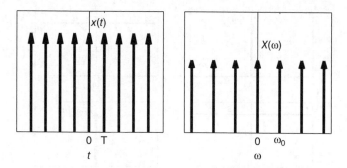

Figure B.11   A train of impulses and the corresponding spectrum

Equation (B.38) shows that the spectrum of a periodic signal is a line spectrum with a density that corresponds to the fundamental frequency, and that the size of the various components (the envelope of the lines) is determined by the corresponding sample values of the Fourier spectrum of one period of the signal, as shown in Figure B.11.

Example B.4   A square wave signal and its Fourier transform: A train of rectangular pulses, a 'square-wave' signal, may be regarded as the convolution of one single rectangular pulse (see Example B.2) and the function given by Equation (B.37). Accordingly, this signal has the spectrum

$$X(\omega) = a\omega_0 \sum_{n=-\infty}^{\infty} t_0 \frac{\sin \frac{n\omega_0 t_0}{2}}{\frac{n\omega_0 t_0}{2}} \delta(\omega - n\omega_0) = a\frac{2\pi t_0}{T} \sum_{n=-\infty}^{\infty} \frac{\sin \frac{n\pi t_0}{T}}{\frac{n\pi t_0}{T}} \delta(\omega - n\omega_0),$$

where t_0 is the width of one pulse. Note that every second line disappears if $t_0 = T/2$ (see also Figure B.12); this signal has only odd-numbered harmonics.

B.4.3   Estimation of the Spectrum of a Periodic Signal

Determining the Fourier spectrum of a periodic signal from a finite record involves time windowing, that is, multiplication with a weighting function, as mentioned in Section B.3.2. From the convolution theorem it follows that the effect of this windowing is that each line of the true line spectrum of the signal is replaced by the spectrum of the time window,

$$x(t)v(t) \leftrightarrow \left(\sum_{n=-\infty}^{\infty} X_p(n\omega_0) \delta(\omega - n\omega_0) \right) * V(\omega) = \omega_0 \sum_{n=-\infty}^{\infty} X_p(n\omega_0) V(\omega - n\omega_0),$$

$$\tag{B.41}$$

where

$$v(t) \leftrightarrow V(\omega). \tag{B.42}$$

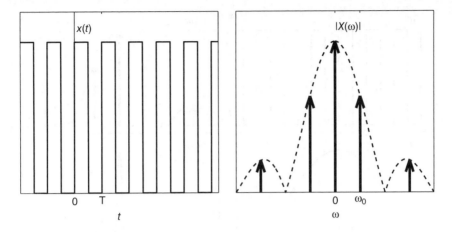

Figure B.12   The spectrum of a sequence of rectangular pulses

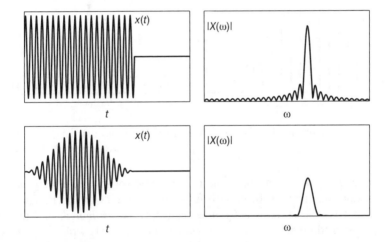

Figure B.13   The effect of the time window on the estimated spectrum of a pure-tone signal. (a) The rectangular window; (b) the Hanning window

Figure B.13 demonstrates the influence of the shape of the time window on the estimated spectrum of a sinusoidal signal. The width of the principal lobe of the spectrum of the Hanning window is broader than that of the rectangular window, but the side lobes decay far more rapidly. Figure B.14 shows the effect of multiplication with the Hanning window on a square-wave signal.

See also the remarks on periodic pseudo-random noise in Section B7.3

B.5   Random Signals

The signals dealt with in Sections B.3 and B.4 are deterministic signals, which means that they can be described mathematically and, accordingly, predicted at all points in

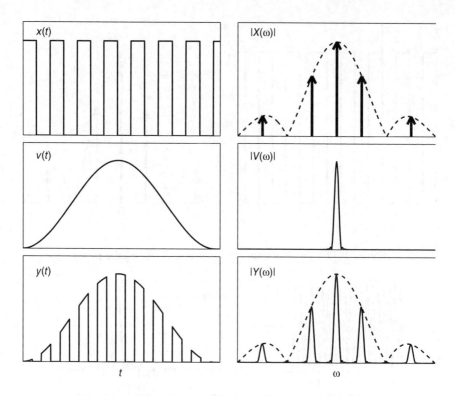

Figure B.14   A Hanning-weighted square-wave signal

time. By contrast, the instantaneous value of a random (or stochastic) signal cannot be predicted. Random signals must be described in terms of their statistical properties. These properties cannot be measured exactly, they can only be *estimated*, and the estimates will be subject to random errors. Only stationary random signals will be considered here. Stationary random signals are signals whose statistical properties do not change with time.

Stationary random signals cannot be Fourier transformed and have no spectrum in the same sense as transients or periodic signals have. The spectral properties of a stationary random signal are described in terms of its *power spectrum*. The power (or mean square value) of the signal is given by the integral

$$P_x = \lim_{T \to \infty} \frac{1}{T} \int_{-T/2}^{T/2} x^2(t) \, dt \tag{B.43}$$

(cf. Equations (B.1) and (B.22)). The power spectrum gives information about the spectral distribution of this power. The square root of P_x is the rms value.

A random signal of considerable practical importance is *Gaussian* (or *normally distributed*) noise. The instantaneous values of this signal have the probability distribution

$$f_x(u) = \frac{1}{\sigma\sqrt{2\pi}} e^{-u^2/2\sigma^2}, \tag{B.44}$$

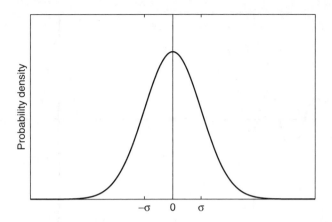

Figure B.15 Probability density function of Gaussian noise

where σ^2 is the variance of the random function $x(t)$ (which here is assumed to have zero mean); see Figure B.15. Since by definition the variance of a random quantity equals the mean square deviation from the average value it follows that the variance of $x(t)$ is identical with the power of the signal, P_{x}. Note that the probability distribution gives no information about the spectral distribution of the noise.

B.5.1 Autocorrelation Functions and Power Spectra

The autocorrelation of a random signal is a function that expresses the statistical relation between the instantaneous values of the signal at two different points in time. It is defined as the average value of the instantaneous product of the signal and a time-shifted version of the signal,

$$R_{xx}(\tau) = E\{x(t)x(t+\tau)\} = \lim_{T \to \infty} \frac{1}{T} \int_{-T/2}^{T/2} x(t)x(t+\tau)\,dt, \qquad (B.45)$$

where $E\{\}$ denotes the average value. Obviously,

$$R_{xx}(0) = P_{x}. \qquad (B.46)$$

It is also evident that the autocorrelation is an even function of τ, and that

$$R_{xx}(0) \geq R_{xx}(\tau). \qquad (B.47)$$

Typical examples of autocorrelation functions are shown in Figure B.16. It is apparent that the autocorrelation function of a random signal gives information about how rapidly the signal changes with time. It is perhaps not surprising, then, that the power spectrum (or power spectral density or autospectrum) is the Fourier transform of the autocorrelation function:

$$S_{xx}(\omega) = \int_{-\infty}^{\infty} R_{xx}(\tau)e^{-j\omega\tau}\,d\tau. \qquad (B.48)$$

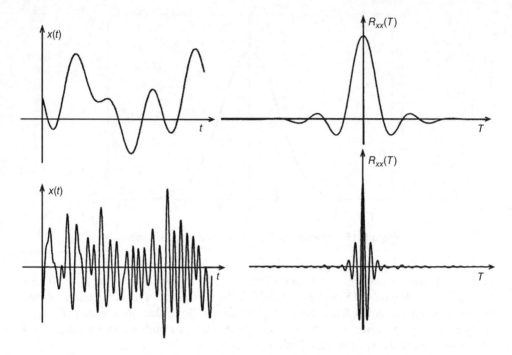

Figure B.16   Examples of random signals and their autocorrelation functions

Since the autocorrelation is an even function of τ, it follows that the power spectrum is a real, even function of ω. From Equations (B.3) and (B.48) we see that

$$P_x = R_{xx}(0) = \frac{1}{2\pi} \int_{-\infty}^{\infty} S_{xx}(\omega)d\omega, \tag{B.49}$$

which shows that the total power of the signal is obtained by integrating the power spectrum over all frequencies (cf. Equation (B.19) and (B.27)).

Example B.5 White noise and its autocorrelation: 'White noise' is a stationary random signal with a flat (frequency-independent) spectral density,

$$S_{xx}(\omega) = A.$$

From Equation (B.33) it follows that the autocorrelation function of white noise is a delta function,

$$R_{xx}(\tau) = A\delta(\tau).$$

It is obvious that this signal, which has infinite power, cannot be realised. However, it is a very useful concept in theoretical work.

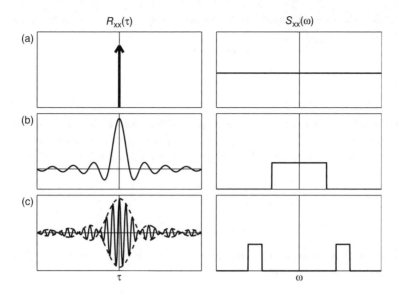

Figure B.17 Autocorrelation functions and power spectra. (a) White noise; (b) bandlimited white noise; (c) bandpass filtered white noise

---

Example B.6 Low-pass filtered white noise and its autocorrelation: Bandlimited noise is noise that has passed through a low-pass or a bandpass filter, a device that attenuates frequencies outside the passband (see Section B.6.1). Figure B.17(b) shows the power spectrum and the corresponding autocorrelation function of white noise filtered by an ideal low-pass filter (cf. Figure B.3). See also Example B.10.

Example B.7 Bandpass filtered white noise and its autocorrelation: White noise passed through an ideal bandpass filter has the autocorrelation function and power spectrum shown in Figure B.17(c). See also Example B.11.

B.5.2 Cross-correlation Functions and Cross-power Spectra

The cross-correlation of two stationary random signals $x(t)$ and $y(t)$ expresses the statistical relation between the two signals. It is defined by the expression

$$R_{xy}(\tau) = E\{x(t)y(t+\tau)\} = \lim_{T \to \infty} \frac{1}{T} \int_{-T/2}^{T/2} x(t)y(t+\tau)\mathrm{d}t. \tag{B.50}$$

The Fourier transform of this function is the cross-power spectrum,

$$S_{xy}(\omega) = \int_{-\infty}^{\infty} R_{xy}(\tau)\mathrm{e}^{-\mathrm{j}\omega\tau}\,\mathrm{d}\tau, \tag{B.51}$$

which, unlike the autospectrum, is a complex function in the general case. From Equation (B.6) we see that

$$S_{xy}(\omega) = S_{xy}^*(-\omega). \tag{B.52}$$

It can also be seen that

$$S_{yx}(\omega) = S_{xy}^*(\omega). \tag{B.53}$$

Example B.8 Cross-correlation between a signal and a delayed version: It is easy to see that cross-correlation functions and cross-spectra are useful in determining the time lag between two identical or closely related random signals; if

$$y(t) = ax(t - t_0)$$

then

$$R_{xy}(\tau) = E\{x(t)ax(t + \tau - t_0)\} = aR_{xx}(\tau - t_0),$$

which assumes its maximum value for $\tau = t_0$, as illustrated in Figure B.18. The corresponding cross-spectrum is

$$S_{xy}(\omega) = a\int_{-\infty}^{\infty} R_{xx}(\tau - t_0)e^{-j\omega\tau}\,d\tau = aS_{xx}(\omega)e^{-j\omega t_0}$$

(cf. Equation (B.11)), which shows that the delay can also be deduced from the slope of the phase of the cross-spectrum. Possible applications include identification of transmission paths and reflections. In room acoustics the most direct approach would be to look at the cross-correlation function. Because of the dispersion of bending waves, the frequency-domain approach is more useful in structural acoustics.

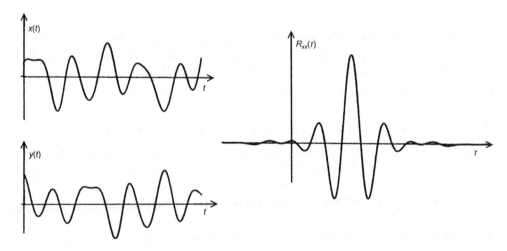

Figure B.18 A random signal, a time-shifted version of the same signal, and the cross-correlation function

Example B.9 A sum of stationary random signals and its autocorrelation: If $z(t)$ is the sum of two stationary random signals,

$$z(t) = x(t) + y(t),$$

then

$$R_{zz}(0) = E\left\{(x(t) + y(t))^2\right\} = R_{xx}(0) + R_{yy}(0) + 2R_{xy}(0),$$

which shows that the power of a sum of two uncorrelated signals (signals whose cross-correlation function is identically zero) is simply the sum of the powers,

$$P_z = P_x + P_y,$$

just as the power of a sum of sinusoids is the sum of the powers of the components. Moreover,

$$R_{zz}(\tau) = R_{xx}(\tau) + R_{yy}(\tau)$$

and

$$S_{zz}(\omega) = S_{xx}(\omega) + S_{yy}(\omega)$$

if the signals are uncorrelated.

The fact that the mean square values of uncorrelated signals are simply added is of great practical importance, because noise signals generated by different mechanisms are invariably uncorrelated.

B.5.3   Estimation of Correlation Functions and Power Spectra

In practice one cannot use an infinite averaging time in determining correlation functions; obviously, they must be estimated from finite segments. It can be seen from Equations (B.17) and (B.45) that the autocorrelation of $x(t)$ bears resemblance to the convolution of $x(t)$ with a time-reversed version of the signal, $x(-t)$. This leads to the conclusion that the autocorrelation of $x(t)$ might be estimated as follows,

$$\tilde{R}_{xx}(\tau) = \frac{1}{T}x_T(t) * x_T(-t), \tag{B.54}$$

where $x_T(t)$ is a segment of the stationary signal $x(t)$ with the duration T. (The caret ˆ indicates an estimate.) The function $\tilde{R}_{xx}(\tau)$ is clearly a *biased* estimate of $R_{xx}(\tau)$ when $\tau \neq 0$, that is

$$E\left\{\tilde{R}_{xx}(\tau)\right\} \neq R_{xx}(\tau). \tag{B.55}$$

It is obvious, for example, that the estimate is zero outside the interval $[-T, T]$. The estimate is, in fact, weighted by a symmetric triangular function of duration $2T$ (the rectangular time window convolved with itself, cf. Example B.3). Whether this weighting is of importance or not depends, of course, on the duration of the true autocorrelation function

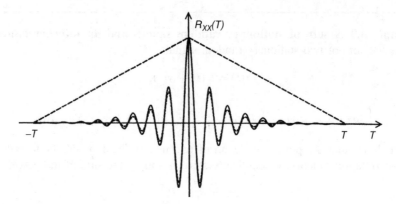

Figure B.19 Autocorrelation function. Solid line, True function; dashed line, weighted estimate; long dashes, weighting function

$R_{xx}(\tau)$; if it decays rapidly compared with T the distortion is negligible. Figure B.19 shows an example of the effect of the weighting function.[5]

The estimate given by Equation (B.55) is not only biased, it is also subject to a random error, simply because it depends on the particulars of the record of $x(t)$ that has been used. However, this error can be reduced to insignificance by averaging sequentially over an appropriate number of records. This is known as 'Welch's method' [3].

The Fourier transform of $\tilde{R}_{xx}(\tau)$ is an estimate of the power spectrum,

$$\tilde{S}_{xx}(\omega) = \int_{-\infty}^{\infty} \tilde{R}_{xx}(\tau) e^{-j\omega\tau}\, d\tau = \frac{1}{T} |X_T(\omega)|^2 \tag{B.56}$$

(cf. Equation (B.17)), where

$$x_T(t) \leftrightarrow X_T(\omega), \tag{B.57}$$

that is, $X_T(\omega)$ is the spectrum of the transient signal $x_T(t)$. It is apparent that the energy spectrum of a sequence of the signal divided by the duration of the sequence is an estimate of the power spectrum. Equation (B.57) shows that the power spectrum of a random signal can be estimated directly from Fourier transformed records of the signal.

Without averaging over a number of records, the estimate $|X_T(\omega)|^2/T$ is a very poor estimate of $S_{xx}(\omega)$ owing to significant random errors, as demonstrated by Figure B.20.[6] Because of the finite duration of the records the estimate is also subject to a resolution bias error, but whether this error is serious or not depends, of course, on the particulars of the signal: the resolution bias error is negligible if the duration of the autocorrelation of the signal is much shorter than T. It should finally be mentioned that a smoother time window than the rectangular function is advantageous, cf. the considerations in Section B.3.2. The Hanning window is very often used.

---

[5] It is possible to compensate for the bias error within the interval $[-T, T]$. Because of its shape, the correction is known as the 'bow-tie correction'.

[6] It can be shown that both the real and the imaginary part of the Fourier transform of the time record are normally distributed, which leads to the conclusion that the relative standard deviation of the estimated power spectrum equals unity [4].

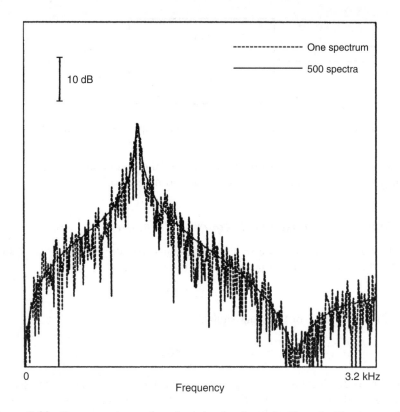

Figure B.20   Power spectrum of random signal estimated with and without averaging

Cross-spectra can be estimated in a similar manner: by averaging sequentially over a number of 'raw' spectral estimates of the form

$$\tilde{S}_{xy}(\omega) = \frac{1}{T} X_T^*(\omega) Y_T(\omega),$$ (B.58)

where

$$y_T(t) \leftrightarrow Y_T(\omega),$$ (B.59)

and the each pair of segments $x_T(t)$ and $y_T(t)$ are recorded simultaneously.

B.6   Linear Systems

One of the characteristics of a dynamic *system* is that an input signal $x(t)$ gives rise to an output signal $y(t)$, as sketched in Figure B.21. For example, the input signal to a loudspeaker in a room may give rise to an output signal from a microphone in the same room. Here we will concentrate on *linear, time-invariant* systems, although weakly nonlinear systems will be touched upon. Linearity implies that if the input signal is a sum of two signals then the output signal equals the sum of the output signals generated by each input signal (linear superposition). The system is time-invariant if a time shift of

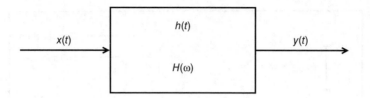

Figure B.21   A linear time-invariant single input/single output system

the input signal results in an identical time shift of the output signal. Most acoustic and mechanical systems may be regarded as linear and time-invariant.

B.6.1   Impulse Response and Frequency Response

If a linear system is driven with a series of transients with the same area but with shorter and shorter duration then the response of the system approaches a limit,

$$y(t) = X(0)h(t),$$

(B.60)

where $X(0)$ is the 'area' of the input signal,

$$X(0) = \int_{-\infty}^{\infty} x(t)\,dt$$

(B.61)

(cf. Equation (B.2)). Note that the response is a transient signal, and that the shape of this signal is independent of the shape of the excitation. For obvious reasons the function $h(t)$ is called the *impulse response* of the system. A linear time-invariant system is completely characterised by its impulse response.

In what follows the input signal $x(t)$ is an *arbitrary* signal of one of the types discussed in Sections B.3, B.4 and B.5. It is reasonable to expect the output signal to belong to the same group of signals. *Any* signal may be regarded as a sum of pulses of short duration, as indicated in Figure B.22. If Δt is sufficiently small then the response of the system will be the corresponding sum of impulse responses. A pulse at $t = t_i$ with the duration Δt and the area $\Delta t \cdot x(t)$ gives rise to the response $\Delta t \cdot x(t_i) \cdot h(t - t_i)$. It follows that the total output signal can be written

$$y(t) = \lim_{\Delta t \to 0} \sum_{i=-\infty}^{\infty} x(t_i)h(t - t_i)\Delta t = \int_{-\infty}^{\infty} x(u)h(t - u)\,du = x(t) * h(t).$$

(B.62)

It is apparent that the output signal of the system is the convolution of the input signal and the impulse response of the signal.

If $x(t)$ is a transient signal or a periodic signal then it follows from the convolution theorem that

$$Y(\omega) = X(\omega)H(\omega),$$

(B.63)

where the Fourier transform of the impulse response,

$$H(\omega) = \int_{-\infty}^{\infty} h(t)e^{-j\omega\tau}\,dt = |H(\omega)|e^{j\varphi(\omega)},$$

(B.64)

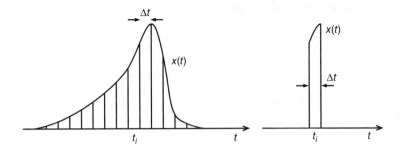

Figure B.22 The input signal divided into a sequence of short pulses

is called the *frequency response* of the system. If $x(t)$ is a sinusoidal signal,

$$x(t) = a \cos \omega_0 t, \tag{B.65}$$

then

$$X(\omega) = \pi a (\delta(\omega - \omega_0) + \delta(\omega + \omega_0)), \tag{B.66}$$

and therefore

$$Y(\omega) = \pi a (\delta(\omega - \omega_0) + \delta(\omega + \omega_0)) H(\omega)$$
$$= \pi a |H(\omega_0)| (\delta(\omega - \omega_0) e^{j\varphi(\omega_0)} + \delta(\omega + \omega_0) e^{-j\varphi(\omega_0)}) H(\omega) \tag{B.67}$$

from which it can be seen that

$$y(t) = a |H(\omega_0)| \cos(\omega_0 t + \varphi(\omega_0)). \tag{B.68}$$

Obviously, the response to a sinusoidal input signal with a certain frequency is a similar signal, but modified in amplitude and shifted in phase by the amplitude response $|H(\omega_0)|$, and the phase response, $\varphi(\omega_0)$, respectively; hence the name frequency response.

Equation (B.62) is also valid when the input signal is a stationary random signal. It can be shown that

$$R_{xy}(\tau) = R_{xx}(\tau) * h(\tau), \tag{B.69}$$

from which it follows that

$$S_{xy}(\omega) = S_{xx}(\omega) \cdot H(\omega). \tag{B.70}$$

It can also be shown that

$$R_{yy}(\tau) = R_{xx}(\tau) * h(\tau) * h(-\tau), \tag{B.71}$$

which corresponds to the equation

$$S_{yy}(\omega) = S_{xx}(\omega) |H(\omega)|^2 = S_{xy}(\omega) H^*(\omega). \tag{B.72}$$

It is obvious that the impulse response of a real system must be zero for $t < 0$: the system must be *causal*; it cannot possibly respond before it is excited. It can be shown [5] that this constraint on $h(t)$ implies that

$$\text{Re}\{H(\omega)\} = \text{Im}\{H(\omega)\} * \frac{1}{\pi\omega}, \tag{B.73a}$$

$$\text{Im}\{H(\omega)\} = -\text{Re}\{H(\omega)\} * \frac{1}{\pi\omega}. \tag{B.73b}$$

The relation between the real and imaginary part of the frequency response is known as the *Hilbert transform*.

Example B.10 An ideal low-pass filter and its impulse response: The ideal low-pass filter is a device with the frequency response

$$H(\omega) = \begin{cases} 1 & \text{if } \omega < |\omega_0| \\ 0 & \text{elsewhere.} \end{cases}$$

From Example B.2 it follows that the impulse response of this filter is the function

$$h(t) = \frac{\omega_0}{\pi} \frac{\sin \omega_0 t}{\omega_0 t}$$

(cf. Figure B.3). Note that this filter is not a causal system: the ideal low-pass filter is conceptually useful, but it cannot be realised. However, it can be approximated arbitrarily well if a sufficient delay is introduced.

Example B.11 An ideal bandpass filter and its impulse response: The ideal bandpass filter has the frequency response

$$H(\omega) = \begin{cases} 1 & \text{if } \omega_c - \Delta\omega/2 < |\omega_0| < \omega_c + \Delta\omega/2 \\ 0 & \text{elsewhere.} \end{cases}$$

where ω_c is the centre frequency and $\Delta\omega$ is the bandwidth of the pass-band. The corresponding impulse response follows from Equation (B.12) and Example B.7:

$$h(t) = \frac{\Delta\omega}{\pi} \frac{\sin \frac{\Delta\omega t}{2}}{\frac{\Delta\omega t}{2}} \cos \omega_c t$$

(cf. Figure B.17). The remarks about causality and practicability in Example B.10 also apply here.

Example B.12 An integrator and its frequency response: An integrator is a device that determines a running average of a signal:

$$y(t) = \frac{1}{T} \int_{t-T}^{t} x(u)\mathrm{d}u.$$

This process is often termed 'linear averaging'. Evidently, it corresponds to convolving with a rectangular pulse with unit area and the duration T, that is, the impulse response is

$$h(t) = \begin{cases} \dfrac{1}{T} & \text{for } 0 < t < T \\ \\ 0 & \text{elsewhere.} \end{cases}$$

From Example B.2 and Equation (B.11) it follows that

$$H(\omega) = \frac{\sin \dfrac{\omega T}{2}}{\dfrac{\omega T}{2}} e^{-j\omega T/2}.$$

This is clearly a low-pass filter with unity gain for $\omega \ll \frac{2\pi}{T}$ and zeros at $\frac{2\pi}{T}, \frac{4\pi}{T}, \frac{6\pi}{T}, \ldots$ cf. Figure B.3. Integrators are used as detectors in combination with square-law devices in rms-measuring instruments.

Example B.13 An exponential detector and its frequency response: Exponential averaging means smoothing with a decaying exponential with unit area,

$$y(t) = \frac{1}{\tau} \int_{-\infty}^{t} x(u) e^{-(t-u)/\tau} \, du.$$

This process is occasionally termed RC-integration because it can be implemented with a simple analogue RC-circuit. The impulse response is

$$h(t) = \begin{cases} \dfrac{1}{\tau} e^{-t/\tau} & \text{for } t > 0 \\ \\ 0 & \text{elsewhere,} \end{cases}$$

and the frequency response is, from Example B.1,

$$H(\omega) = \frac{1}{1 + j\omega\tau},$$

cf. Figure B.2. Exponential averaging is often used in combination with square-law devices in rms-measuring instruments as an alternative to linear averaging. If the signal is bandlimited white noise then a measurement with an exponential detector with a time constant of τ has the same statistical uncertainty as a measurement with linear averaging over a period of 2τ [6]; hence the quantity 2τ is sometimes referred to as 'the equivalent integration time'.

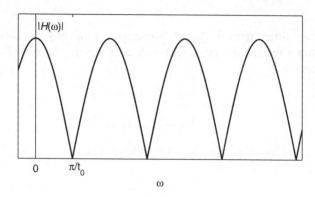

Figure B.23 The frequency response of a comb filter.

Example B.14 A comb filter and its frequency response: A comb filter is a device with the impulse response

$$h(t) = a(\delta(t) + \delta(t - t_0)).$$

The corresponding frequency response is

$$H(\omega) = \int_{-\infty}^{\infty} a(\delta(t) + \delta(t - t_0)) e^{-j\omega t}\, dt = a(1 + e^{-j\omega t_0}) = 2a \cos\left(\frac{\omega t_0}{2}\right) e^{j\omega t_0 / 2};$$

see Figure B.23. Note that the frequency response is periodic with zeros at $\omega = \pi/t_0, 3\pi/t_0, 5\pi/t_0, \ldots$ Comb filter effects occur whenever a delayed version of a signal is added to the signal. For example, a listener placed in front of a reflecting wall will hear sound from a frontal source through a comb filter, because of the interference between direct and reflected sound.

B.6.2 Estimation of the Frequency Response of a Linear System

Measurement of frequency response functions is one of the most important applications of signal analysis in acoustics. There are many methods of measuring the frequency response of a linear system. A classic but outdated technique is to drive the system with a sinusoidal signal with a frequency that is slowly swept through the range of interest. This method relies on Equation (B.68) and involves measuring the amplitude and phase (relative to the excitation) of the response. Equation (B.68) has been derived under the assumption that the input signal is a stationary pure-tone signal. This means that the frequency sweep should be so slow that quasi-stationary conditions are maintained; therefore the acceptable sweeping speed depends on the system under test. Because of the sharp peaks and troughs a *very* slow sweep is required in measuring the response of a lightly damped system. A modern version of this method uses a 'stepped sine'. The analyser moves on to the next tone when the system has reached steady state (this is checked by repeated measurements).

The most common method in acoustics (except room acoustics) is probably the method based on excitation with stationary random noise. From Equation (B.70) it follows that

$$H(\omega) = \frac{S_{xy}(\omega)}{S_{xx}(\omega)}. \tag{B.74}$$

Alternatively, it can be seen from Equation (B.72) that

$$H(\omega) = \frac{S_{yy}(\omega)}{S_{xy}^*(\omega)}. \tag{B.75}$$

Incidentally, Equations (B.74) and (B.75) show that

$$|S_{xy}(\omega)|^2 = S_{xx}(\omega)S_{yy}(\omega). \tag{B.76}$$

In practice we rarely have a pure single input/single output system: the available signals are usually more or less contaminated by uncorrelated random noise from trans-ducers, electronic circuits etc., as indicated in Figure B.24. If $x(t)$ and $y(t)$ are the noise-contaminated, accessible signals,

$$x(t) = a(t) + m(t), \tag{B.77a}$$

$$y(t) = b(t) + n(t), \tag{B.77b}$$

where $m(t)$ and $n(t)$ are extraneous, uncorrelated noise signals and $a(t)$ and $b(t)$ are the true, inaccessible input/output signals,

$$b(t) = h(t) * a(t), \tag{B.78}$$

then

$$S_{xx}(\omega) = S_{aa}(\omega) + S_{mm}(\omega), \tag{B.79a}$$

$$S_{xy}(\omega) = S_{ab}(\omega), \tag{B.79b}$$

$$S_{yy}(\omega) = S_{bb}(\omega) + S_{nn}(\omega), \tag{B.79c}$$

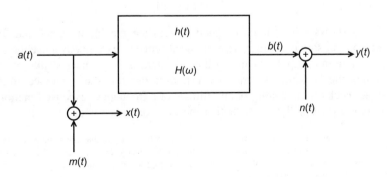

Figure B.24 A linear system with extraneous noise

that is, the extraneous noise signals have no influence on the cross-spectrum (cf. Example B.9). It can now be seen that

$$\tilde{H}(\omega) = \frac{S_{xy}(\omega)}{S_{xx}(\omega)} = \frac{S_{ab}(\omega)}{S_{aa}(\omega) + S_{mm}(\omega)} = H(\omega)\frac{S_{aa}(\omega)}{S_{aa}(\omega) + S_{mm}(\omega)}, \qquad (B.80)$$

which demonstrates that extraneous *output* noise has no systematic influence[7] on the estimate given by Equation (B.74). However, it is obvious that extraneous input noise gives rise to an error. By contrast, the alternative estimator (Equation (B.75)) is affected systematically only by extraneous *output* noise:

$$\tilde{H}'(\omega) = \frac{S_{yy}(\omega)}{S_{xy}^*(\omega)} = \frac{S_{bb}(\omega) + S_{nn}(\omega)}{S_{ab}^*(\omega)} = H(\omega)\frac{S_{bb}(\omega) + S_{nn}(\omega)}{S_{bb}(\omega)}. \qquad (B.81)$$

In general, extraneous noise is more likely to be a problem at the output than at the input since one usually has control over the input signal; therefore the estimator[8] given by Equation (B.74) generally gives the best estimate. However, this may not be the case at the resonances of the system under test, where the response is particularly strong. Figure B.25(a) demonstrates the performance of the two estimators in the presence of strong output noise. It is apparent that they are in agreement in a frequency range about the resonance of the system, but it can be seen that the alternative estimator H' overestimates very significantly at frequencies outside this range.

It is worth noting that Equation (B.76) is no longer valid when there is extraneous noise. The normalised squared magnitude of the cross spectrum, that is, the quantity

$$\gamma_{xy}^2(\omega) = \frac{|S_{xy}(\omega)|^2}{S_{xx}(\omega)S_{yy}(\omega)}, \qquad (B.82)$$

is called the *coherence* (or the ordinary coherence as opposed to the complex coherence) of $x(t)$ and $y(t)$. From Equations (B.80) and (B.81) it can be seen that

$$\gamma_{xy}^2 = \frac{\tilde{H}(\omega)}{\tilde{H}'(\omega)} = \frac{S_{aa}(\omega)S_{bb}(\omega)}{(S_{aa}(\omega) + S_{mm}(\omega))(S_{bb}(\omega) + S_{nn}(\omega))}. \qquad (B.83)$$

Obviously,

$$0 \leq \gamma_{xy}^2(\omega) \leq 1. \qquad (B.84)$$

Figure B.25(b) shows the coherence function between the input signal and the noise-contaminated output signal of the system. In the absence of extraneous noise the coherence between the input and output signal of a linear system is unity, irrespective of the nature of the system or the shape of its frequency response; therefore the coherence function provides a useful check on the quality of an estimate of a frequency response function: changing the excitation level will reveal whether the low coherence is due to extraneous noise.

---

[7] Extraneous noise increases the *random* error of the estimate, which means that the number of 'raw' spectral estimates that are needed in estimating the power and cross-power spectra reliably must be increased; cf. Section B.5.3. In the absence of extraneous noise the random error is negligible, even though the random errors of $S_{xx}(\omega)$ and $S_{xy}(\omega)$ are not. See, e.g., Reference [4].

[8] The two estimators (Equations (B.80) and (B.81)) are sometimes referred to as H_1 and H_2, respectively.

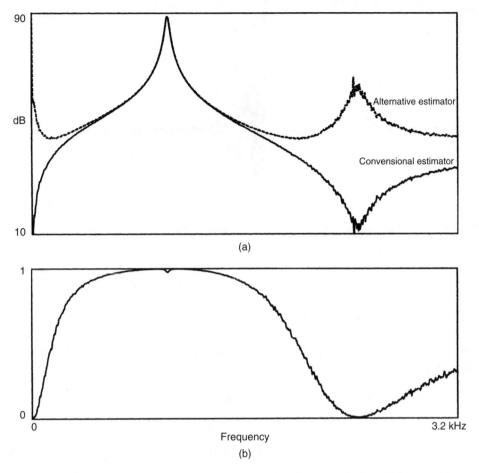

Figure B.25 (a) The frequency response of a system estimated with two different estimators. The output signal is contaminated by uncorrelated white noise. Solid line, Conventional estimator (Equation (B.74)); dashed line, alternative estimator (Equation (B.75)). (b) The corresponding coherence function

Yet another, somewhat more subtle, source of error in estimation of frequency responses is due to resolution bias errors in the estimates of power and cross-power spectra that enter into Equations (B.74) and (B.75). From the considerations in Section B.5.3 it can be concluded that the distortion of the estimated cross-spectrum corresponds to multiplying the cross-correlation function with the self-convolved time window, as sketched in Figure B.26. Inspection of Equation (B.69) suggests that the duration of the cross-correlation function in general exceeds the duration of the impulse response of the system, which leads to the conclusion that significant errors occur when the impulse response of the system under test is longer than the record length T used in the analysis. This 'truncation error' is also reflected in the coherence function: if the impulse response of the system

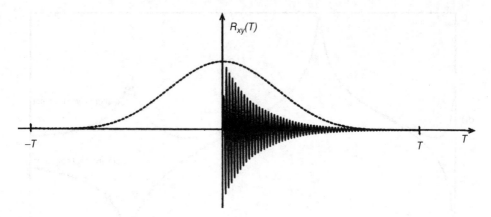

Figure B.26 The cross-correlation function weighted by the Hanning window convolved with itself

is truncated by the estimation procedure it seems as if the output signal is only partly due to the input signal (even in the absence of extraneous output noise), and as a result, the coherence is underestimated. It is a clear indication of estimation errors if the measured coherence function depends on the record length of the analysis. Figure B.27 demonstrates this effect. Another indication of possible truncation errors can be obtained by changing the excitation to synchronised pseudo-random noise, as described in section B.7.3.

Frequency response functions can also be estimated using transient excitation. In principle, one could simply excite the system with a delta function; with

$$x(t) = a\delta(t), \tag{B.85}$$

$$H(\omega) = Y(\omega)/a \tag{B.86}$$

(in the absence of extraneous noise). However, in practice it will rarely be possible to excite the system with a pulse with a flat Fourier spectrum, and therefore one must compensate for the shape of the input spectrum:

$$H(\omega) = \frac{Y(\omega)}{X(\omega)} \tag{B.87}$$

(cf. Equation (B.63)).

One of the problems in using transient excitation is that the energy of a short transient is limited; it may be difficult to obtain sufficient energy in the excitation and in the response in the entire frequency range of interest without overloading the system or violating its linearity. This means that extraneous output noise is likely to be a problem in a measurement where the system is excited with one single transient. However, the effect of extraneous output noise can be reduced by averaging, as follows. Let the signal

$$y_i(t) = x_i(t) * h(t) + n_i(t) \tag{B.88}$$

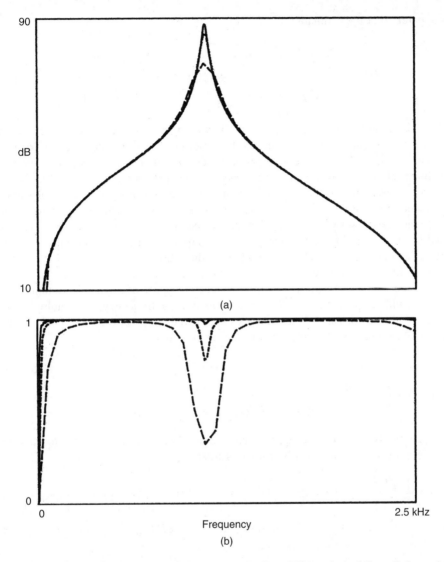

Figure B.27 The influence of the record length: Solid line, 250 ms; dashed line, 62.5 ms; dotted line, 15.6 ms. (a) Estimated frequency response; (b) estimated coherence between input and output signal

be the noise-contaminated response of the system to the i'th transient, where the second term is the part of the stationary noise signal $n(t)$ that is transmitted through the time window. (The time window is chosen so as to have a negligible influence on the response.) It follows that

$$Y_i(\omega) = X_i(\omega)H(\omega) + N_i(\omega). \tag{B.89}$$

It is apparent that the estimate

$$\tilde{H}(\omega) = \frac{\frac{1}{m}\sum\limits_{i=1}^{m} X_i^*(\omega)Y_i(\omega)}{\frac{1}{m}\sum\limits_{i=1}^{m} |X_i(\omega)|^2} = H(\omega) + \frac{\frac{1}{m}\sum\limits_{i=1}^{m} X_i^*(\omega)N_i(\omega)}{\frac{1}{m}\sum\limits_{i=1}^{m} |X_i(\omega)|^2} \tag{B.90}$$

will approach the true frequency response function as m increases, since the phase spectrum of the different noise sequences can be expected to vary randomly from one record to another. This and other averaging procedures for suppressing background noise are sometimes referred to as *signal enhancement*. Note that there is no need for the input transients to be identical; therefore the method can be used for instance in measurement of frequency responses (and thus impulse responses) in rooms with exploding paperbags or frequency responses of mechanical structures based on excitation with hammer blows.

Fast linear or logarithmic frequency sweeps may be regarded as transients, and unlike short pulses they have a moderate crest factor (the ratio of peak to rms value). They are often used in room acoustics and in audio engineering; see, e.g., References [7, 8].

Yet another method of determining the frequency response of a dynamic system involves driving it with a 'stepped sine' (that is, frequency by frequency) and multiplying the response with the corresponding complex exponential (in practice, two signals, a cosine and a sine). A linear system responds to the input signal

$$x(t) = a\cos\omega_0 t \tag{B.91}$$

with the output signal

$$y(t) = a|H(\omega_0)|\cos(\omega_0 t + \varphi(\omega_0)) \tag{B.92}$$

(cf. Equation (B.68)). If the output signal is multiplied with $2e^{-j\omega_0 t}$ then the frequency response at ω_0 can be obtained by low-pass filtering:

$$\lim_{T\to\infty} \frac{1}{T}\int_{-T/2}^{T/2} y(t)2e^{-j\omega_0 t}\,dt = \lim_{T\to\infty} \frac{1}{T}\int_{-T/2}^{T/2} 2a|H(\omega_0)|\cos(\omega_0 t + \varphi(\omega_0))e^{-j\omega_0 t}\,dt$$

$$= a|H(\omega_0)|e^{j\varphi(\omega_0)} = aH(\omega_0) \tag{B93}$$

B.6.3 Estimation of the Frequency Response of a Weakly Nonlinear System

Strictly, the concept of a frequency response is no longer valid in case of a nonlinear system. However, some methods can give meaningful results for weakly nonlinear systems. For example, the 'stepped sine' method can also be used when analysing weakly nonlinear systems (such as, e.g., loudspeakers), and it may be of interest to analyse the distortion. A nonlinear system will not only respond at the frequency of the sine but also at multiples of the excitation frequency. The second harmonic, for example, can be determined by multiplying with $2e^{-2jw_0 t}$ instead of $2e^{-jw_0 t}$.

Frequency sweeps can also be used for analysing weakly nonlinear systems. For example, in loudspeaker measurements under anechoic conditions it is advantageous to use excitation with a logarithmic frequency sweep. Since the distortion products (harmonics) generated by the nonlinear device under test sweep faster through the frequency range than the excitation signal they show up in the measured impulse response *before* the direct sound arrives (as so-called 'harmonic impulse responses'), and thus it should be possible to measure the frequency responses corresponding to the fundamental and the harmonics [7]. However, it has recently been shown that some distortion artifacts do contaminate the impulse response [8]; and thus it cannot be recommended to drive the loudspeakers hard.

B.7 Digital Signal Processing

Many years ago, in the late 1960s and early 1970s, digital analysis techniques replaced analogue methods. All the analysis and estimation procedures referred to in the foregoing are in practice realised with digital signal processing, which implies that the physical, continuous, analogue signals from, say, microphones, particle velocity transducers, accelerometers and force transducers, are replaced by sequences of numbers.

The purpose of this chapter is just to give a *very* brief introduction to digital analysis of signals.

B.7.1 Sampling

A digital (or discrete) signal is a sequence of numbers. To be analysed with digital means, the analogue, continuous signal that directly represents a physical quantity is sampled equidistantly. The result of sampling the analogue signal $x(t)$ periodically with the frequency $\omega_s = 2\pi/\Delta T$ is the signal

$$x_s(t) = x(t)\Delta T \sum_{m=-\infty}^{\infty} \delta(t - m\Delta T) = \Delta T \sum_{m=-\infty}^{\infty} x(m\Delta T)\delta(t - m\Delta T). \qquad (B.94)$$

The corresponding Fourier spectrum follows from Equation (B.16) and (B.37):

$$X_s(\omega) = X(\omega) * \sum_{n=-\infty}^{\infty} \delta(\omega - n\omega_s) = \sum_{n=-\infty}^{\infty} X(\omega - n\omega_s). \qquad (B.95)$$

Equations (B.94) and (B.95) are illustrated in Figure B.28. Note that the spectrum is a periodic function with period ω_s. It is apparent that spectral overlap is avoided if $x(t)$ is a bandlimited signal, provided that the sampling frequency ω_s is sufficiently high: if

$$X(\omega) = 0 \quad \text{for} \quad |\omega| < \omega_s/2 \qquad (B.96)$$

then

$$X_s(\omega) = X(\omega) \quad \text{for} \quad |\omega| < \omega_s/2. \qquad (B.97)$$

Thus the *sampling theorem* states that a bandlimited signal can be represented in terms of its sample values without any loss of information provided that it does not contain frequency components above half the sampling frequency (the so-called Nyquist frequency) [5]. One cannot distinguish between a signal component at ω_0 and one at $\omega_s - \omega_0$

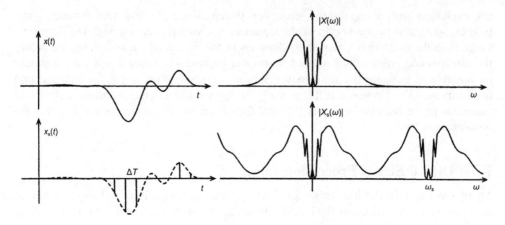

Figure B.28   Illustration of the sampling theorem

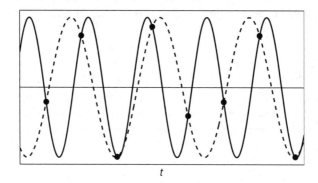

Figure B.29   Sampling of a pure-tone signal

from the sample values, as illustrated in Figure B.29. This phenomenon is called 'aliasing', and the low-pass filter that ensures that the signal to be sampled is bandlimited is called an *anti-aliasing* filter.

B.7.2   The Discrete Fourier Transform

It will be recalled that the Fourier transform of a periodic signal is a sequence of equidistant pulses (cf. Equation (B.38)). Conversely, we have just seen that the Fourier transform of a sequence of equidistant pulses is a periodic function (cf. Equation (B.95)). The *discrete* Fourier transform relates a sampled time signal to a sampled spectrum, and because of the sampling, both functions are now implicitly periodic.

The discrete Fourier transform of the sequence $c_0, c_1, \ldots, c_{N-1}$ is the sequence $C_0, C_1, \ldots, C_{N-1}$,

$$C_n = \frac{1}{N} \sum_{m=0}^{N-1} c_m e^{-jmn\frac{2\pi}{N}}. \tag{B.98}$$

The correspondence is unique. The inverse transform is

$$c_m = \sum_{n=0}^{N-1} C_n e^{jmn\frac{2\pi}{N}}. \tag{B.99}$$

From Equation (B.94) it follows that the periodically extended spectrum $X(\omega)$ can be written as

$$\sum_{i=-\infty}^{\infty} X(\omega - i\omega_s) = \Delta T \int_{-\infty}^{\infty} \sum_{m=0}^{N-1} x(m\Delta T)\delta(t - m\Delta T)e^{-j\omega t}\,dt$$

$$= \Delta T \sum_{m=0}^{N-1} x(m\Delta T)e^{-j\omega m \Delta T}. \tag{B.100}$$

With $\omega = n\Delta\omega$, where

$$\Delta\omega = \frac{\omega_s}{N} = \frac{2\pi}{N\Delta T} = \frac{2\pi}{T} \tag{B.101}$$

is the spectral resolution, Equation (B.100) becomes

$$\sum_{i=-\infty}^{\infty} X(n\Delta\omega - i\omega_s) = \Delta T \sum_{m=0}^{N-1} x(m\Delta T)e^{-jmn\frac{2\pi}{N}}, \tag{B.102}$$

which can be identified as the discrete Fourier transform of N sample values of $x(t)$ (apart from a factor of $N\Delta T = T$). The frequency shifted spectra centred at $\pm\omega_s, \pm 2\omega_s, \ldots$, have no influence if the signal has been sampled in accordance with the sampling theorem; therefore Equation (B.102) demonstrates that the discrete Fourier transform of N sample values of a signal provides N sample values of the Fourier spectrum of the signal. The sampling of the spectrum corresponds to a periodic repetition of the time function.

Because it deals with sequences of numbers, the discrete Fourier transform is far more suitable for computation than the Fourier integral. It can be calculated using an extremely efficient algorithm known as the 'Fast Fourier Transform', FFT [9], which, however, restricts N to certain values, usually a power of two, say, 1024 or 65576.

B.7.3 Signal Analysis with the 'Fast Fourier Transform' (FFT)

There are several limitations of spectral analysis based on the discrete Fourier transform: i) the signal must be bandlimited, and the sampling frequency must be at least twice the highest frequency component; ii) the analysis is based on segments of duration $T = N\Delta T$, which means that spectral estimates are affected by the choice of the time window[9]; iii) only $N/2$ sample values of the spectrum (corresponding to the positive frequencies) are available[10]; and iv) because of the periodic nature of the transform, multiplication in one domain corresponds to *circular* convolution in the other domain.

---

[9] Window effects, which correspond to spectral smoothing, are often referred to as 'leakage': apparently power from a discrete frequency component 'leaks' into adjacent bands.

[10] The sampling of the continuous spectrum is occasionally referred to as the 'picket fence' effect: the continuous function is viewed through the slits of a picket fence, and some details might be missing.

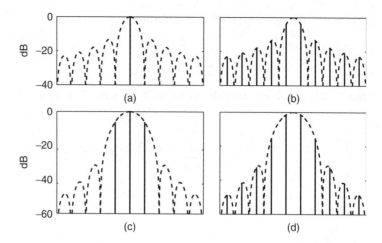

Figure B.30 Sampling of the continuous spectrum of a time-windowed sinusoidal. (a) The rectangular window; tone at sample point; (b) the rectangular window, tone midway between two sample points; (c) the Hanning window, tone at sample point; (d) the Hanning window, tone midway between two sample points

The fact that the analysis effectively corresponds to determining sample values of the Fourier transform of a periodic repetition of the signal means that it is important that any discontinuity between the two ends of the record is smoothed. This is one of the reasons why the Hanning window is often used in analysing stationary signals. In the very special case where the signal is periodic with a fundamental frequency that is multiple of $\Delta\omega$, so that T equals a multiple of the period of the signal, this is not necessary, and the rectangular window of duration T (that is, no weighting function) is the best choice. In this case there is no estimation error. However, unless one has control over the signal the Hanning window is a far better choice.

Figure B.30, which shows the estimated spectrum of a single sinusoid, demonstrates the combined effect of the time window and the spectral sampling. It is apparent that the zeros of the spectrum of the rectangular window coincide with the sample points when the frequency of the pure tone is a multiple of $\Delta\omega$ (which corresponds to the record length being a multiple of the period of the sinusoidal); therefore only one spectral component is detected in this case. The worst case occurs when the frequency of the analysed signal is midway between two adjacent spectral sample points; in this case a number of 'false' frequency components occur. Moreover, the detected maximum value is too low: $-3.9\,\mathrm{dB}$ with the rectangular window and $-1.4\,\mathrm{dB}$ with the Hanning window. This phenomenon is called pass-band ripple.[11]

Power and cross-power spectra of random signals are determined as described in Section B.5.3. Thus, by averaging the squared modulus $|X(N\,\Delta\omega)|^2$ over a number of segments an estimate of the power spectrum of $x(t)$ is obtained. Note that each 'spectral

---

[11] A particular time window called 'flat top' is often used for calibration purposes where amplitude errors in pure-tone measurements cannot be tolerated. The adjective 'flat' refers to the response in the frequency domain: this window has negligible pass-band ripple [10].

line' is in effect a bandpass filter with a bandwidth of the order of $\Delta\omega$.[12] It follows that one can detect a periodic signal buried in noise by analysing with fine spectral resolution: as the resolution becomes finer less and less noise will pass through each filter and the discrete frequency components of the periodic signal will emerge.

Correlation functions can be estimated by calculating the inverse discrete Fourier transform of the corresponding power and cross-power spectral estimates. Because of the efficiency of the FFT algorithm, this is much faster than a direct calculation of the convolution (Equation (B.70)). However, since the multiplications that enter into the spectral estimates correspond to circular convolution (convolution of periodically extended functions), the results are correct only for small values of the time shift ($\tau \ll T$). The solution to this problem is called 'zero padding': no window is applied, but the second half of each time record is set to zero. With zero padding the correlation functions will be weighted by a triangular function as described in Section B.5.3.

A multi-channel FFT analyser averages corresponding values of $|X(N\Delta\omega)|^2$, $|Y(N\Delta\omega)|^2$ and $X^*(N\Delta\omega)Y(N\Delta\omega)$ (if we restrict ourselves to two channels), from which estimates of $S_{xx}(\omega)$, $S_{yy}(\omega)$ and $S_{xy}(\omega)$ are determined. In turn the frequency response between the two signals can be calculated, either from Equation (B.74) or Equation (B.75), and the coherence function can be calculated from Equation (B.82). The impulse response can be calculated as the inverse Fourier transform of the frequency response. All these quantities are, of course, affected by the smoothing that derives from the time window. In particular, an error occurs if the impulse response of the system under test is longer than the duration of the time records, T; cf. Figure B.27. Insufficient spectral resolution (too short time records) are reflected in poor coherence. Thus the number of spectral lines in the FFT analysis should be increased until no further improvement in the coherence is seen.

Multi-channel FFT analysers are often used with random noise excitation, but other excitation signals are possible. For example, some FFT analysers can generate synchronised 'pseudo-random noise', which is a periodic signal with frequency components that correspond exactly to the spectral lines of the analyser [12]. Because the different frequency components have random phase relations, the signal resembles (and sounds like) random noise. However, if such a signal is analysed using the rectangular window, the result will be correct sample values of the spectrum, since each frequency component coincides with a sample point. In other words, there is no leakage. It follows that one can determine *correct sample values* of the frequency response of a linear system with this method irrespective of the spectral resolution; therefore the estimated coherence is unity irrespective of the frequency resolution. However, this is no guarantee of a correct measurement; if the frequency response varies rapidly with the frequency, it may well be undersampled (one may miss a peak or a trough). If the function is undersampled an aliasing error will occur in the *time* domain. This corresponds to the impulse response being longer than the time record of the analysis, and the result is not a truncation error but a wrap-around error.

Sometimes the system under test may not be completely linear. If the system is weakly nonlinear (as, e.g., loudspeakers) the dual channel FFT method with random noise will give the best linear approximation. This is not the case if the excitation is pseudo-random noise, because all the frequency components in the periodic signal will produce intermodulation distortion [10].

---

[12] The bandwidth depends on the duration and shape of the time window [11].

B.7.4   The Method Based on 'Maximum Length Sequences' (MLS)

With the advent of cheap and powerful digital signal processing systems many different methods of measuring frequency and impulse responses have become available, one of which shall be briefly described here.

In the 1980s a correlation technique based on excitation with so-called maximum length sequences, became popular, in particular in audio engineering and in room acoustics [13, 14]. A maximum-length sequence is a binary sequence whose circular autocorrelation (except for a small DC-error) is a delta function. In the MLS-method the system under test is driven with pseudo-random noise consisting of a periodically repeated maximum length sequence, and the (circular) cross-correlation between this signal and the output signal is calculated using an extremely efficient algorithm known as the fast Hadamard transform, FHT [15]. (This algorithm restricts the length of the sequence to certain values: a power of two minus one.) Since the autocorrelation of the excitation is a delta function, the cross-correlation is proportional to the impulse response of the system (cf. Equations (B.45) and (B.70)). Note that the MLS-method is a time domain method; the direct result is an estimate of the impulse response. If the frequency response is wanted it must be calculated subsequently with an FFT-transform.

If the signal-to-noise ratio is large so that no averaging is required, the MLS-method is faster than the conventional dual channel FFT method based on excitation with random noise, and because of the efficiency of the transform one can easily cope with sequences of hundreds of thousands of samples, which makes it realistic to determine the impulse response of a room with a sampling frequency that covers the entire audible frequency range. When the MLS-method was developed no FFT analyser in commercial production could do that, and FFT analysers are still not practical for this application. However, the fact that the Hadamard transform is computationally even more efficient than the FFT transform is obviously much less important now than it was in the 1980s.

It is apparent that there are strong similarities between this method and the FFT technique based on the use of synchronised pseudo-random noise: in both cases a wrap-around error will occur if the impulse response of the system under test is longer than the period of the signal. However, this is avoided if the sequence is long enough. On the other hand, the use of very long sequences can be problematic: the longer the period of the excitation, the more sensitive is the method to minute changes of the system, for example because of small temperature fluctuations.[13] Obviously the problem is aggravated if it is necessary to average over many sequences. Unfortunately one cannot determine the coherence with the MLS-method, so there is no indication of such problems.[14]

The MLS method is not without problems. Nonlinearities give rise to serious errors, for the same reason as mentioned above in the description of FFT measurements with pseudo-random noise. Another disadvantage of the MLS method is due to the fact that it does not determine the impulse response of the system under test, it determines the impulse response of the *total* system between the generator signal and the output of the

---

[13] Such fluctuations clearly violate the assumption of time invariance, and this is very serious for all methods based on pseudo-random sequences derived from number theory [16].

[14] An MLS coherence function with properties similar to the usual coherence can be determined if averaging over several periods of the signal is carried out [17]. Since the MLS signal is periodic this measure cannot detect nonlinearities (because the distortion will be the same in all periods), but it can indicate the presence of background noise or a system that is not time-invariant.

system, including filters and transducers. Whereas the two-channel FFT method based on Equations (B.74) or Equation (B.75) automatically compensates for the shape of the spectrum of the 'true excitation', the MLS-method does not. Figure B.31 demonstrates the problem. An electrodynamic exciter driven with white noise excites an undamped plate; the force signal is measured with a force transducer, and the response of the plate is measured with a small accelerometer. Figure B.31(a) shows that the spectrum of the force signal is far from flat, and Figure B.31(b) shows the frequency response of the structure and, for comparison, the frequency response between the electrical signal that drives the exciter (the voltage) and the response measured on the structure. Note how the force signal is reduced at resonances.

One solution to this problem is to 'prewhiten' or 'pre-emphasise' the signal from an initial measurement so as to make the spectrum of the actual excitation flat. Within some limits this may work in room acoustics and in audio, but it makes the method inapplicable to, say, structural dynamics, since the mechanical impedance of most structures varies so strongly with the frequency (in a manner that differs from point to point) that one simply cannot generate a frequency independent force signal (cf. Figure B.31). Moreover, attempts to do this by shaping the spectrum of the input signal to an exciter that drives a weakly damped structure are bound to generate severe distortion. A better solution to the

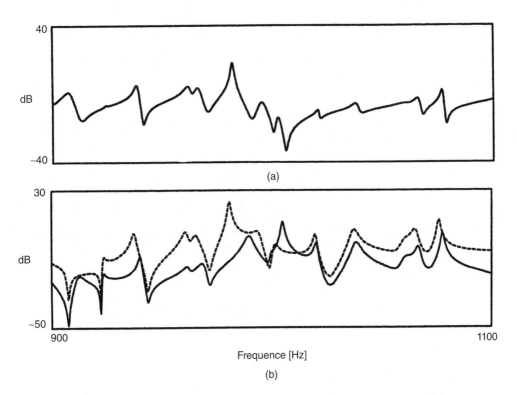

(a)

Frequence [Hz]

(b)

Figure B.31   (a) Power spectrum of force signal; (b) Solid line, frequency response of the structure under test, that is, frequency response between force signal and accelerometer signal; dashed line, frequency response of total system, including the exciter that drives the structure

problem caused by the fact that the MLS method is fundamentally a single channel method is to measure *twice* and thus determine the frequency response between the generator signal and the output of the system under test and the frequency response between the generator signal and the 'true excitation' (e.g., the force). The frequency response of the system is then the ratio of these two functions.

References

[1] A. Papoulis: *The Fourier Integral and Its Applications*. McGraw-Hill, New York (1962).

[2] M.J. Lighthill: *An Introduction to Fourier Analysis and Generalized Functions*. Cambridge University Press (1964).

[3] P.D. Welch: The use of fast transform for the estimation of power spectra: A method based on time averaging over short, modified periodograms. *IEEE Transactions on Audio and Electroacoustics* **15**, 70–73 (1967).

[4] J.S. Bendat: Statistical errors in measurement of coherence and unput/output quantities. *Journal of Sound and Vibration* **59**, 405–421 (1978).

[5] A.Papoulis: *Signal Analysis*. McGraw-Hill, New York (1977). See Chapters 5 and 7.

[6] J. Pope: Analyzers. Chapter 107 in *Handbook of Acoustics*, ed. M.J. Crocker. John Wiley & Sons, New York (1998).

[7] S. Müller and P. Massarani: Transfer-function measurements with sweeps. *Journal of the Audio Engineering Society* **49**, 443–471 (2004).

[8] A. Torras Rosell and F. Jacobsen: A new interpretation of distortion artifacts in sweep measurements. *Journal of the Audio Engineering Society* **59**, 283–289 (2011).

[9] J.W. Cooley and J.W. Tukey: An algorithm for the machine calculation of complex Fourier series. *Mathematics of Computation* **19** (90), 297–301 (1965).

[10] S. Gade and H. Herlufsen: Use of weighting functions in DFT/FFT analysis (Parts I and II). *Brüel & Kjær Technical Review* **3** and **4** (1987).

[11] F.J. Harris: On the use of windows for harmonic analysis with the discrete Fourier transform. *Proceedings of IEEE* **66**, 51–83 (1978).

[12] H. Herlufsen: Dual channel FFT analysis (Parts I and II). *Brüel & Kjær Technical Review 1 and 2* (1984).

[13] D.D. Rife and J. Vanderkooy: Transfer-function measurement with maximum-length sequences. *Journal of Audio Engineering Society* **37**, 419–443 (1989).

[14] W.T. Chu: Impulse-response and reverberation-decay measurements made by using a periodic pseudo-random sequence. *Applied Acoustics* **29**, 193–205 (1990).

[15] J. Borish and J.B. Angell: An efficient algorithm for measuring the impulse response using pseudorandom noise. *Journal of Audio Engineering Society* **31**, 478–487 (1983).

[16] A. Torras Rosell and F. Jacobsen: Measuring long impulse responses with pseudorandom sequences and sweep signals. *Proceedings of Inter-Noise 2010*, Lisbon, Portugal (2010).

[17] J. Liu and F. Jacobsen: An MLS coherence function and its performance in measurements on time-varying systems. *Proceedings of Sixth International Congress on Sound and Vibration*, pp. 2851–2858, Copenhagen, Denmark (1999).

Appendix C

Cylindrical and Spherical Bessel Functions; Legendre Functions; and Expansion Coefficients

C.1 Cylindrical Bessel Functions

The solutions to the differential equation

$$\frac{d^2 y}{dz^2} - \frac{1}{z}\frac{dy}{dx} + \left(1 - \frac{m^2}{z^2}\right)y = 0 \qquad m = 0, 1, 2, \ldots,\tag{C.1}$$

where m is an integer, are the cylindrical Bessel functions of the first and second kind

$$y(z) = A\mathrm{J}_m(z) + B\mathrm{N}_m(z).\tag{C.2}$$

The cylindrical Bessel functions of the first and second kind are convenient for describing the sound field in terms of standing waves–i.e. for interior problems. For exterior problems two other linearly independent functions that satisfy Equation (C.1) can easily be constructed,

$$\mathrm{H}_m^{(1)}(z) = \mathrm{J}_m(z) + \mathrm{j}\mathrm{N}_m(z),\tag{C.3}$$

$$\mathrm{H}_m^{(2)}(z) = \mathrm{J}_m(z) - \mathrm{j}\mathrm{N}_m(z).\tag{C.4}$$

These are the cylindrical Bessel functions of the third kind. The cylindrical Bessel functions of the second kind are also termed the cylindrical Neumann functions, and the cylindrical Bessel functions of the third kind are also called the cylindrical Hankel functions.

The Bessel function can be expressed as a series,

$$\mathrm{J}_m(z) = \left(\frac{z}{2}\right)^m \sum_{n=0}^{\infty} \frac{(-1)^n}{n!(m+n)!}\left(\frac{z}{2}\right)^{2n},\tag{C.5}$$

Fundamentals of General Linear Acoustics, First Edition. Finn Jacobsen and Peter Møller Juhl.
© 2013 John Wiley & Sons, Ltd. Published 2013 by John Wiley & Sons, Ltd.

from which expressions for small arguments can be found,

$$J_0(z) = 1 + O(z^2),$$

$$J_1(z) = z/2 + O(z^3),$$

$$J_m(z) = \frac{z^m}{2^m m!} + O(z^{m+2}).$$

(C.6a-c)

The corresponding series expansion for the Neumann function is more complicated, but for small arguments

$$N_0(z) \simeq \frac{2}{\pi} \left(\ln \frac{z}{2} + \gamma \right),$$

(C.7a)

$$N_m(z) \simeq -\frac{(m-1)!}{\pi} \left(\frac{2}{z} \right)^m,$$

(C.7b)

is found, where $\gamma = 0.57721...$ is Euler's constant.

For large arguments the Bessel functions' relations with the ordinary trigonometric functions become apparent since,

$$\lim_{z \to \infty} J_m(z) = \sqrt{\frac{2}{\pi z}} \cos(z - m\pi/2 - \pi/4),$$

(C.8)

and

$$\lim_{z \to \infty} N_m(z) = \sqrt{\frac{2}{\pi z}} \sin(z - m\pi/2 - \pi/4).$$

(C.9)

Some useful relations between Bessel functions are

$$J_{-m}(z) = (-1)^m J_m(z),$$

(C.10a)

$$N_{-m}(z) = (-1)^m N_m(z),$$

(C.10b)

$$J_{m-1} + J_{m+1} = \frac{2m}{z} J_m(z),$$

(C.11)

and

$$J_{m-1} - J_{m+1} = 2\frac{dJ_m(z)}{dz}.$$

(C.12)

For $m = 0$ Equation (C.12) specialises to

$$\frac{dJ_0(z)}{dz} = -J_1(z),$$

(C.13)

by the use of Equation (C.10a).

Combining Equations (C.10) to (C.13) results in the following rules, which are often used,

$$\frac{dJ_m}{dz} = J_{m-1} - \frac{m}{z} J_m = -J_{m+1} + \frac{m}{z} J_m,$$

$$\int J_1(z) dz = -J_0(z),$$

$$\int z J_0(z) dz = z J_1(z), \tag{C.14a-e}$$

$$\int z J_0^2(z) dz = \frac{z^2}{2} (J_0^2(z) + J_1^2(z)),$$

$$J_0(z) + J_2(z) = \frac{2}{z} J_1(z).$$

Note that Equations (C.11) to (C.14a-e) have analogue versions for the Neumann and Hankel functions, N_m and H_m.

For small arguments the Hankel function is dominated by the Neumann function, and combining Equations (C.7) and (C.12) results in

$$\frac{d H_m^{(2)}(z)}{dz} \rightarrow -\frac{jm!}{\pi \epsilon_m} \left(\frac{2}{z} \right)^{m+1}, \quad \text{for} \quad z \rightarrow 0, \tag{C.15}$$

where $\epsilon_0 \equiv 1$ and $\epsilon_m \equiv 2$ for $m \geq 1$.

The Bessel function may also be defined as an integral,

$$J_m(z) = \frac{j^{-m}}{\pi} \int_0^\pi e^{jz \cos \theta} \cos(m\theta) d\theta. \tag{C.16}$$

Several other interesting relations between Bessel functions, their derivatives and anti-derivatives exist. The reader is referred to standard books on mathematics, e.g. Reference [1].

C.2   Legendre Functions

The solutions to the differential equation

$$(1 - x^2) \frac{d^2 y}{dx^2} - 2x \frac{dy}{dx} + m(m+1) y = 0 \qquad m = 0, 1, 2, \ldots \tag{C.17}$$

are the Legendre polynomials $P_m(x)$,

$$P_m(x) = \frac{1}{2^m m!} \left[\frac{d^m}{dx^m} (x^2 - 1)^m \right]. \tag{C.18}$$

For our purpose the substitution $x = \cos \theta$ is used, which transforms equation (C.17) to

$$\frac{d^2 y}{d\theta^2} + \frac{\cos \theta}{\sin \theta} \frac{dy}{d\theta} + m(m+1) y = 0. \tag{C.19}$$

The first few Legendre polynomials are given below

$$
\begin{array}{ll}
P_0(x) = 1 & P_0(\cos \theta) = 1 \\
P_1(x) = x & P_1(\cos \theta) = \cos \theta \\
P_2(x) = \frac{1}{2}(3x^2 - 1) & P_2(\cos \theta) = \frac{1}{4}(3 \cos(2\theta) + 1) \\
P_3(x) = \frac{1}{2}(5x^3 - 3x) & P_3(\cos \theta) = \frac{1}{8}(5 \cos(3\theta) + 3 \cos \theta) \\
\vdots & \vdots
\end{array}
$$

An important feature of the Legendre polynomials is their orthogonality,

$$\int_{-1}^{1} P_m(x)P_n(x)dx = \frac{2\delta_{mn}}{2m+1},$$ (C.20)

or

$$\int_{0}^{\pi} P_m(\cos\theta)P_n(\cos\theta)\sin\theta d\theta = \frac{2\delta_{mn}}{2m+1}.$$ (C.21)

Several other relations between Legendre functions, their derivatives and anti-derivative exist – the reader is referred to e.g. Reference [1], from where the following useful result is taken:

$$\int_{-1}^{1} x^m P_m(x)dx = 2\int_{0}^{1} x^m P_m(x)dx = \frac{2^{m+1}(m!)^2}{(2m+1)!}.$$ (C.22)

C.3 Spherical Bessel Functions

The solutions to the differential equation

$$\frac{d^2y}{dz^2} - \frac{2}{z}\frac{dy}{dz} + \left(1 - \frac{m(m-1)}{z^2}\right)y = 0 \qquad m = 0, 1, 2,...,$$ (C.23)

are the spherical Bessel functions of the first and second kind $j_m(z)$ and $n_m(z)$.[1] Two other linearly independent functions that satisfy equation (C.23) can easily be constructed,

$$h_m^{(1)}(z) = j_m(z) + jn_m(z),$$ (C.24)

$$h_m^{(2)}(z) = j_m(z) - jn_m(z).$$ (C.25)

These are the spherical Bessel functions of the third kind. The spherical Bessel functions of the second kind are also termed the spherical Neumann functions, and the spherical Bessel functions of the third kind are also called the spherical Hankel functions.

Often mathematical packages include only the cylindrical Bessel functions (J, N and H). However, the spherical functions are easily obtained from the cylindrical, since the relation

$$j_m(z) = \sqrt{\frac{\pi}{2z}}J_{m+1/2}(z)$$ (C.26)

holds for all Bessel functions (j_m, n_m and h_m).

The spherical Bessel functions are closely related to the normal trigonometric functions:

$$j_0(z) = \frac{\sin z}{z}$$

$$j_1(z) = \frac{\sin z}{z^2} - \frac{\cos z}{z}$$

$$j_2(z) = \left(\frac{3}{z^3} - \frac{1}{z}\right)\sin z - \frac{3}{z^2}\cos z$$ (C.27)

$$\vdots$$

---

[1] In what follows the spherical Bessel function of the first kind always has an index m, which may be used to distinguish it from the imaginary unit.

and

$$n_0(z) = -j_{-1}(z) = -\frac{\cos z}{z}$$

$$n_1(z) = j_{-2}(z) = -\frac{\cos z}{z^2} - \frac{\sin z}{z}$$

$$n_2(z) = -j_{-3}(z) = \left(-\frac{3}{z^3} + \frac{1}{z}\right)\cos z - \frac{3}{z^2}\sin z \qquad (C.28)$$

$$\vdots$$

An important feature of all Bessel functions (j, n and h) is that their derivatives are related to a Bessel function of another order,

$$m j_{m-1} - (m+1)j_{m+1} = (2m+1)\frac{dj_m(z)}{dz}. \qquad (C.29)$$

Often the far-field limit $z \to \infty$ is of interest:

$$\lim_{z\to\infty} j_m(z) = \frac{1}{z}\cos(z - (n+1)\pi/2)$$

$$\lim_{z\to\infty} n_m(z) = \frac{1}{z}\sin(z - (n+1)\pi/2)$$

$$\lim_{z\to\infty} h_m^{(1)}(z) = \frac{1}{z}e^{j(z-(n+1)\pi/2)} \qquad (C.30a\text{-}d)$$

$$\lim_{z\to\infty} h_m^{(2)}(z) = \frac{1}{z}e^{-j(z-(n+1)\pi/2)}$$

as well as the limits of small arguments $z \to 0$,

$$\lim_{z\to0} j_m(z) = \frac{z^m}{1\cdot3\cdot5\cdots(2m+1)} = \frac{z^m 2^m m!}{(2m+1)!} \qquad (C.31a)$$

$$\lim_{z\to0} n_m(z) = \frac{1\cdot3\cdot5\cdots(2m-1)}{z^{m+1}} \qquad (C.31b)$$

Several other interesting relations between Bessel functions, their derivatives and anti-derivatives exist. The reader is referred to standard books on mathematics, e.g. Reference [1].

C.4   Expansion Coefficients

At the surface of a sphere a plane wave may be expanded in spherical coordinates,

$$\hat{p}_{inc} = e^{-jkr\cos\theta}e^{j\omega t}$$

$$= \sum_{m=0}^{\infty} B_m j_m(kr)P_m(\cos\theta)e^{j\omega t}. \qquad (C.32)$$

In order to find the expansion coefficients B_m Equation (C.32) is multiplied with P_n and integrated over the spherical surface. Making use of Equation (C.21)

$$\int_0^\pi e^{-jkr\cos\theta} P_n(\cos\theta) \sin\theta \, d\theta = \frac{2B_n}{2n+1} j_n(kr) \tag{C.33}$$

is obtained, which by change of variables ($x = \cos\theta$) becomes

$$\int_{-1}^1 e^{-jkrx} P_n(x) \, dx = \frac{2B_n}{2n+1} j_n(kr). \tag{C.34}$$

Equation (C.34) must be valid at all r and at $r = 0$ in particular. In this special case both sides of Equation (C.34) are zero unless $n = 0$ (since P_n is orthogonal to $P_0 = 1$ and since $j_n(0) = 0$ for $n > 0$). Thereby $B_0 = 1$ is easily found.

For $n > 0$ the calculations are somewhat more tedious. Equation (C.34) is differentiated n times with respect to r and then evaluated at $r = 0$. Since

$$\frac{d^n}{dr^n}(e^{-jkrx}) = (-jkx)^n (e^{-jkrx}), \tag{C.35}$$

the left hand side becomes (at $r = 0$)

$$(-jk)^n \int_{-1}^1 x^n P_n(x) \, dx = (-jk)^n \frac{2^{n+1}(n!)^2}{(2n+1)!}, \tag{C.36}$$

where Equation (C.22) has been used in the last result.

Now consider the right hand side of Equation (C.34). From Equation (C.31a) it follows that

$$\frac{d^n}{dr^n}(j_n(kr))\bigg|_{r=0} = k^n \frac{2^n(n!)^2}{(2n+1)!}, \tag{C.37}$$

so that the nth derivative of the right hand side of Equation (C.34) becomes (at $r = 0$)

$$\frac{d^n}{dr^n}\left(\frac{2B_n}{2n+1} j_n(kr)\right)\bigg|_{r=0} = k^n \frac{B_n}{2n+1} \frac{2^{n+1}(n!)^2}{(2n+1)!}. \tag{C.38}$$

Comparing Equations (C.36) and (C.38) gives

$$B_n = (-j)^n (2n+1). \tag{C.39}$$

Reference

[1] M. Abramowitz and I.A. Stegun: *Handbook of Mathematical Functions*. National Bureau of Standards, Washington, D.C., 1964.

Appendix D

Fundamentals of Probability and Random Variables

D.1 Random Variables

Random (or stochastic) variables can be described in terms of their probability density functions. By definition the integral of such a function from minus infinity to infinity is unity,

$$P\{-\infty < x < \infty\} = \int_{-\infty}^{\infty} f_x(u)\mathrm{d}u = 1, \tag{D.1}$$

indicating that the variable must always assume some value. The integral of the probability density over a certain interval gives the probability of the variable being in that interval. Thus

$$P\{a \leq x \leq b\} = \int_{a}^{b} f_x(u)\mathrm{d}u \tag{D.2}$$

is the probability that the random variable x with the probability density f_x takes a value in the interval from a to b.

It is sometimes sufficient to characterise a random variable in terms of its average (or mean or expected) value and its variance or standard deviation. The average value of x is

$$E\{x\} = \int_{-\infty}^{\infty} u f_x(u)\mathrm{d}u. \tag{D.3}$$

The variance is

$$\sigma^2\{x\} = E\{(x - E\{x\})^2\} = \int_{-\infty}^{\infty} (u - E\{x\})^2 f_x(u)\mathrm{d}u. \tag{D.4}$$

The square root of the variance, σ, is called the standard deviation. It is often useful to normalise the standard deviation with the average value,

$$\varepsilon\{x\} = \sigma\{x\}/E\{x\}. \tag{D.5}$$

Fundamentals of General Linear Acoustics, First Edition. Finn Jacobsen and Peter Møller Juhl.
© 2013 John Wiley & Sons, Ltd. Published 2013 by John Wiley & Sons, Ltd.

D.2   The Central Limit Theorem

The central limit theorem states that a sum of independent random variables having the same probability distribution tends to become normally distributed, that is, the probability function of the sum

$$z = x_1 + x_2 + \cdots + x_n \tag{D.6}$$

tends to

$$f_z(u) \simeq \frac{1}{\sigma \sqrt{2\pi}} e^{-(u-\mu)^2/2\sigma^2}, \tag{D.7}$$

where

$$\mu = n \times E\{x_i\} \tag{D.8}$$

and

$$\sigma = \sqrt{n \times \sigma^2\{x_i\}}. \tag{D.9}$$

Actually it is not even necessary for the independent random variables to have the same probability distribution for the sum to become normally distributed; see [1].

D.3   Chi and Chi-Square Statistics

Let $x_1, x_2, x_3, \ldots, x_n$ be independent normally distributed (or Gaussian) variables with zero mean, variance σ^2 and the probability density

$$f_{x_i}(u) = \frac{1}{\sigma \sqrt{2\pi}} \exp\left(-\frac{u^2}{2\sigma^2}\right), \tag{D.10}$$

corresponding to the joint probability function

$$f(u_1, u_2, \ldots, u_n) = \frac{1}{(\sigma \sqrt{2\pi})^n} \exp\left(-\frac{u_1^2 + u_2^2 + \cdots + u_n^2}{2\sigma^2}\right). \tag{D.11}$$

We now define two random variables,

$$\chi_n = \sqrt{x_1^2 + x_2^2 + \ldots + x_n^2} \tag{D.12}$$

and

$$\chi_n^2 = x_1^2 + x_2^2 + \ldots + x_n^2. \tag{D.13}$$

They are known as chi and chi-square statistics with n degrees of freedom, respectively, and it can be shown that they have the following probability density functions [1],

$$f_{\chi_n}(u) = \begin{cases} 0 & \text{for } u < 0 \\ \dfrac{2}{\left(\sqrt{2}\sigma\right)^n \Gamma(n/2)} u^{n-1} e^{-u^2/2\sigma^2} & \text{for } u \geq 0, \end{cases} \tag{D.14}$$

$$f_{\chi_n^2}(u) = \begin{cases} 0 & \text{for } u < 0 \\ \dfrac{u^{n/2-1}}{\left(\sqrt{2}\sigma\right)^n \Gamma(n/2)} e^{-u/2\sigma^2} & \text{for } u \geq 0, \end{cases} \tag{D.15}$$

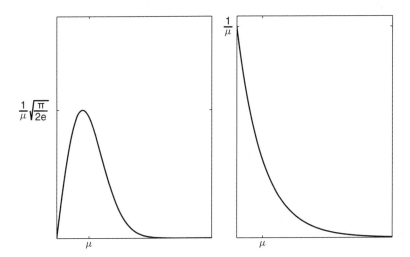

Figure D.1 Left: Rayleigh distribution; right: exponential distribution. In both cases the parameter is the mean value

where Γ is the gamma function,

$$\Gamma(z) = \int_0^\infty t^{z-1} e^{-t} dt. \tag{D.16}$$

If the argument of the gamma function is a positive integer m, then

$$\Gamma(m) = (m-1)! = 1 \times 2 \times \cdots \times (m-1). \tag{D.17}$$

In the particular case where $n = 2$ Equations (D.14) and (D.15) become the Rayleigh probability density,

$$f_{\chi_2}(u) = \begin{cases} 0 & \text{for } u < 0 \\ \dfrac{u}{\sigma^2} e^{-u^2/2\sigma^2} & \text{for } u \geq 0, \end{cases} \tag{D.18}$$

and the exponential probability density,

$$f_{\chi_2^2}(u) = \begin{cases} 0 & \text{for } u < 0 \\ \dfrac{e^{-u/2\sigma^2}}{2\sigma^2} & \text{for } u \geq 0, \end{cases} \tag{D.19}$$

respectively; see Figure D.1. It is easy to show that a Rayleigh distributed random variable has a relative standard deviation of

$$\varepsilon\{\chi_2\} = \sqrt{\frac{4-\pi}{\pi}} \simeq 0.52, \tag{D.20}$$

whereas an exponentially distributed variable has a relative standard deviation of unity:

$$\varepsilon\{\chi_2^2\} = 1. \tag{D.21}$$

Reference

[1] A. Papoulis and S.P. Pillai: *Probability, Random Variables, and Stochastic Processes* (4th ed.). McGraw-Hill, New York (1991).

Bibliography

L. Cremer and H. Müller: *Principles and Applications of Room Acoustics* (Volumes 1 and 2). Applied Science Publishers Ltd, London (1982)

M.J. Crocker (editor): *Encyclopedia of Acoustics*. John Wiley & Sons, New York (1997)

F.J. Fahy: *Sound Intensity* (2nd edition). E & FN Spon, London (1995)

F.J. Fahy: *Foundations of Engineering Acoustics*. Academic Press, San Diego (2001)

H. Kuttruff: *Room Acoustics* (4th edition). E & FN Spon, London (2000)

P.M. Morse: *Vibration and Sound* (2nd edition). The American Institute of Physics (1983)

P.M. Morse and K.U. Ingard: *Theoretical Acoustics*. Princeton University Press (1984)

M.L. Munjal: *Acoustics of Ducts and Mufflers*. John Wiley & Sons, New York (1987)

A. Papoulis: *Signal Analysis*. McGraw-Hill, New York (1977)

A. Papoulis and S.P. Pillai: *Probability, Random Variables, and Stochastic Processes* (4th edition). McGraw-Hill, New York (1991).

A.D. Pierce: *Acoustics: An Introduction to Its Physical Principles and Applications*. The American Institute of Physics (1989)

E.G. Williams: *Fourier Acoustics: Sound Radiation and Nearfield Acoustical Holography* Academic Press, San Diego (1999)

Fundamentals of General Linear Acoustics, First Edition. Finn Jacobsen and Peter Møller Juhl.
© 2013 John Wiley & Sons, Ltd. Published 2013 by John Wiley & Sons, Ltd.

Index

Note: Page numbers in *italics* refer to Figures; those in **bold** to Tables

Fundamentals of General Linear Acoustics, First Edition. Finn Jacobsen and Peter Møller Juhl.
© 2013 John Wiley & Sons, Ltd. Published 2013 by John Wiley & Sons, Ltd.